Climate Change:
Financing Global Forests

Climate Change:
Financing Global Forests

THE ELIASCH REVIEW

publishing for a sustainable future

London • Sterling, VA

First published by Earthscan in the UK and USA in 2008

© Crown copyright 2008

The text of the Eliasch Review reproduced in this document may be used free of charge in any format or medium for research, private study or internal circulation within an organisation providing it is reproduced accurately and not used in a misleading context. The material must be acknowledged as Crown copyright and the title of the document specified.

The Report and supporting material can be found on the Office of Climate Change website at www.occ.gov.uk.

Any enquiries relating to the copyright in this document should be addressed to:
The Information Policy team, HMSO, St Clements House, 2–16 Colegate, Norwich, NR3 1BQ
email: licensing@opsi.x.gsi.gov.uk.

All rights reserved

ISBN 978-1-84407-772-4 hardback
      978-1-84407-773-1 paperback

Printed and bound in the UK by MPG Press, Bodmin
Cover design by Susanne Harris

For a full list of publications please contact:

Earthscan
Dunstan House
14a St Cross St
London EC1N 8XA
UK
tel: +44 (0)20 7841 1930
fax: +44 (0)20 7242 1474
email: earthinfo@earthscan.co.uk
Web: www.earthscan.co.uk

22883 Quicksilver Drive, Sterling, VA 20166-2012, USA

Earthscan publishes in association with the International Institute for Environment and Development

A catalogue record for this book is available from the British Library

Library of Congress Cataloging-in-Publication Data

Eliasch, Johan.
  Climate change : financing global forests : the Eliasch review.
     p. cm.
  "Commissioned by the Prime Minister, the Review is an independent report to government, prepared by Johan Eliasch with the support of the Office of Climate Change"
  Includes bibliographical references and index.
  ISBN 978-1-84407-772-4 (hardback) -- ISBN 978-1-84407-773-1 (pbk.) 1. Forest protection--Finance. 2. Sustainable forestry--International cooperation. 3. Deforestation--Prevention. 4. Climatic changes--Prevention. I. Title. II. Title: Eliasch review.
  SD411.E45 2008
  333.75'137--dc22   2008046936

The paper used for this book is FSC-certified.
FSC (the Forest Stewardship Council) is an
international network to promote responsible
management of the world's forests.

Climate Change: Financing Global Forests

The Eliasch Review

Table of contents

Preface	ix
Background papers	xi
Acknowledgements	xiii
Executive summary	xv
1. Introduction	**1**
1.1 The impacts of climate change	2
1.2 Climate change mitigation	5
1.3 Forests and climate change	6
1.4 Forest communities and ecosystem services	8
1.5 The scope of this Review	10

Part I: The challenge of deforestation

2. Forests, climate change and the global economy	**15**
2.1 Forests and the carbon cycle	16
2.2 Impacts of human activities on the forest carbon cycle	18
2.3 Impacts of forests on climate change	23
2.4 Modelling future impacts	26
2.5 Conclusion	33
3. The drivers of deforestation	**35**
3.1 Why are trees being cut down?	36
3.2 Population growth and wealth creation	37
3.3 Growing demand for agricultural products and timber	39
3.4 Current economic incentives for landowners to deforest	41
3.5 Policy incentives	42
3.6 Land tenure	44
3.7 Capacity	45
3.8 Forest transitions over time	47
3.9 Conclusion	48

4.	**Sustainable production and poverty reduction**	49
4.1	Introduction	50
4.2	Land availability	50
4.3	A vision of sustainable production	52
4.4	Sustainable production and conservation	53
4.5	Infrastructure and alternative employment	58
4.6	Forest conservation	60
4.7	Key levers for shifting to more sustainable production	62
4.8	Conclusion	68
5.	**The costs of mitigation**	69
5.1	Introduction	70
5.2	Up-front and ongoing mitigation costs	70
5.3	Ongoing forest emissions reduction costs	71
5.4	Estimating the opportunity costs of avoided deforestation	72
5.5	Estimating the costs of purchasing forest emissions abatement	75
5.6	The benefits of taking action to reduce forest emissions	77
5.7	Conclusion	80

Part II: Forests and the international climate change framework: the long-term goal

6.	**A long-term framework for tackling climate change**	83
6.1	Overall framework for tackling climate change	84
6.2	Criteria for a successful climate change framework	85
6.3	Comparison of options for achieving global climate stabilisation	90
6.4	Rationale for including forests within a global cap and trade system	95
6.5	Four key elements of a long-term framework	98
6.6	Conclusion	99
7.	**The current international climate change framework**	101
7.1	Current international action	102
7.2	The United Nations Rio Conventions	102
7.3	The importance of the Kyoto Protocol	107
7.4	Limitations of the first Kyoto commitment period	111
7.5	Bali Action Plan	117
7.6	Conclusion	117

Part III: The building blocks of forest financing: the medium-term approach

8.	**Transition to a long-term framework**	**121**
8.1	Introduction	122
8.2	Types of transition path	123
8.3	A three-stage transition process: short, medium and long term	125
8.4	Conclusion	127
9.	**Effective targets for reducing forest emissions**	**129**
9.1	Introduction	130
9.2	Baseline level	130
9.3	Determining the baseline	133
9.4	Baseline trajectories	141
9.5	Conclusion	143
10.	**Measuring and monitoring emissions from forests**	**145**
10.1	The importance of robust measuring and monitoring	146
10.2	Measuring carbon stocks in forests	147
10.3	Monitoring and verifying emissions and sequestration	155
10.4	International and national approaches to measuring and monitoring	159
10.5	Capacity building: expertise and costs	162
10.6	Conclusion	164
11.	**Linking to carbon markets**	**165**
11.1	Introduction	167
11.2	Carbon markets: supply and demand	168
11.3	Price impacts of linking forest credits to emissions trading schemes	174
11.4	Scale of carbon market finance for forest abatement	182
11.5	Linking mechanism	184
11.6	Conclusion	189
12.	**Governance and distribution of finance**	**191**
12.1	Introduction	192
12.2	National-level governance	192
12.3	Distribution of finance	196
12.4	International governance	205
12.5	Conclusion	210

Part IV: International action, capacity building and short-term funding

13.	**The funding gap and capacity building**	**213**
13.1	Introduction	214
13.2	Research, analysis and knowledge sharing	214
13.3	Policy and institutional reform	216
13.4	Demonstration activities	219
13.5	Meeting the funding gap	222
13.6	Coordination and governance of public funding	229
13.7	Conclusion	232
14.	**Conclusions**	**233**
14.1	Introduction	234
14.2	The forest sector in a global climate change deal	234
14.3	International cooperation to support capacity building	236
14.4	Coordinated international action to deliver finance effectively	237
14.5	Conclusion	238

Bibliography	**241**
Index	**251**

Preface

This Review was commissioned by the Prime Minister. The Review is an independent report to government, prepared by Johan Eliasch with the support of the Office of Climate Change. It aims to provide a comprehensive analysis of international financing to reduce forest loss and its associated impacts on climate change. It does so with particular reference to the international efforts to achieve a new global climate change agreement in Copenhagen at the end of 2009.

The Review focuses on the scale of finance required to produce significant reductions in forest carbon emissions, and the mechanisms that, if designed well, can achieve this effectively to help meet a global climate stabilisation target. It also examines how mechanisms to address forest loss can contribute to poverty reduction, as well as the importance of preserving other ecosystem services such as biodiversity and water services.

Approach to the Review

A range of new research and analysis was commissioned for this Review from the following international organisations and institutes:

- AEA
- Chatham House
- Climate Strategies
- CSERGE, University of East Anglia
- EcoSecurities
- International Energy Solutions (IES)
- International Institute for Applied Systems Analysis (IIASA)
- International Institute for Environment and Development (IIED)
- Judge Business School, University of Cambridge
- LTS International
- The Met Office Hadley Centre
- Overseas Development Institute (ODI)
- ProForest
- The Royal Botanic Gardens, Kew
- School of Biological Sciences, Plymouth University
- United Nations Environment Programme World Conservation Monitoring Centre (UNEP/WCMC)

This Review also draws on a large amount of previous research in the literature. The subject of carbon finance and global forests is complex and wide-ranging, and not all literature, particularly in some of the specialised subject areas, could be cited in this report. However, where more information is sought on any section of this Review, we recommend referring to the previously published reviews, summary articles and more detailed references that are cited in the report.

During the Review, the team visited a number of countries to learn from projects and policies on the ground in forest nations, including Brazil, Indonesia, Cameroon and the Democratic Republic of Congo. The team also met with representatives of Papua New Guinea and Guyana.

In preparing its analysis the team has consulted broadly. Submissions to the Review were invited in March 2008 and a series of meetings and round-tables were held in May 2008 with representatives from NGOs, academic institutions and business groups who responded to the questionnaire. These included Fauna and Flora International, Greenpeace, The Rainforest Foundation, Sustainable Forest Management Ltd, The Prince's Rainforests Project, Department for International Development, Global Canopy Programme, Forests Philanthropy Action Network, University of Leicester, Centre for Environmental Research, Quest, University of Reading, Forestry Commission, WWF, Down to Earth and Global Witness

About the author

Johan Eliasch is the Prime Minister's Special Representative on Deforestation and Clean Energy. In this role he was commissioned to undertake an independent review on the role of international finance mechanisms to reduce the loss of global forests in tackling climate change.

A team from the Office of Climate Change (OCC) supported Johan Eliasch in conducting the Review and acted as its secretariat. The OCC works across HM Government to support analytical work on climate change and the development of climate change policy and strategy.

The executive summary, full report, background papers and further information are available from www.occ.gov.uk.

Background papers

A series of background papers was produced based on the research and analysis commissioned for the Review from international academics, experts, organisations and institutes. These papers were used as part of the evidence-gathering process to inform the Review:

- Baalman, P and Schlamadinger, B (2008) *Scaling up AFOLU Mitigation Activities in Non-Annex I Countries*, Climate Strategies

- Betts, R et al (2008) *Forests and emissions: a contribution to the Eliasch Review*, The Met Office Hadley Centre

- Franco, M (2008) *Carbon absorption and storage*, School of Biological Sciences, Plymouth University

- Grieg-Gran, M (2008) *Costs of avoided deforestation*, International Institute for Environment and Development (IIED)

- Gusti, M et al (2008) *Technical Model of the IIASA model cluster*, International Institute for Applied Systems Analysis (IIASA)

- Hoare et al (2008) *Estimating the cost of building capacity in rainforest nations to allow them to participate in a global REDD mechanism*, Chatham House, ProForest, Overseas Development Institute (ODI), EcoSecurities

- Hope, C (2008) *Valuing the climate change impacts of tropical deforestation*, Judge Business School, University of Cambridge

- Hope, C and Castilla-Rubio, J C (2008) *A first cost benefit analysis of action to reduce deforestation*, Judge Business School, University of Cambridge

- Moat, J et al (2008) *Rapid forest inventory and mapping: Monitoring forest cover and land use change*, The Royal Botanic Gardens, Kew

- Miles, L et al (2008) *Mapping vulnerability of tropical forest to conversion and resulting potential CO_2 emissions*, UNEP/WCMC

- Hardcastle, P et al (2008) *Capability and cost assessment of the major forest nations to measure and monitor their forest carbon*, LTS International

- Sajwaj, T et al (2008) *The Eliasch Review: Forest management impacts on ecosystem services*, AEA

- Sathaye, J et al (2008) *Updating carbon density and opportunity cost parameters in deforesting regions in the GCOMAP model*, International Energy Solutions (IES)

Further details of the background papers and the approach of the Review are available on the OCC website at www.occ.gov.uk.

Acknowledgements

The team in the Office of Climate Change was led by Graham Floater. Team members included Joanna Follett, Colin Mackie, Michael Mullan, Duncan Stone, Jenny Ward and Judith Whiteley. They were supported by Nana Osei-Asibey, Chris Westrop, Nat Martin, James Vause, Rachel Lewis and Jo O'Driscoll.

The following individuals and organisations provided background papers that were used to inform the Review: Maryanne Grieg-Gran (International Institute for Environment and Development); Mykola Gusti, Petr Havlik and Michael Obersteiner (International Institute for Applied Systems Analysis); Jayant Sathaye, Peter Chan, Helcio Blum, Larry Dale and Willy Makundi (IES); Alison Hoare, Thomas Legge, Duncan Brack (Chatham House); Ruth Nussbaum (Proforest); Jade Saunders (EFI/Chatham House); David Brown (ODI); Justin Moat, Charlotte Crouch, William Milliken, Paul Smith, Martin Hamilton and Susan Baena (Royal Botanic Gardens, Kew); Pat Hardcastle and Dr David Baird (LTS), with Virginia Harden and contributions from P G Abbot (LTS), P O'Hara (LTS), J R Palmer (Forest Management Trust), Andy Roby (DFID Indonesia), T Haüsler (GAF AG), Vitus Ambia (PNG), Anne Branthomme (FAO), Mette Wilkie (FAO), Ernesto Arends (Venezuela) and Carlos González (Mexico); Richard Betts, Jemma Gornall, John Hughes, Neil Kaye, Doug McNeall and Andy Wiltshire (Met Office Hadley Centre); Lera Miles, Valerie Kapos, Igor Lysenko and Alison Campbell (UNEP World Conservation Monitoring Centre) with contributions from Holly Gibbs (Center for Sustainability and the Global Environment, University of Wisconsin, Madison); Todd Sajwaj, Mike Harley and Clare Parker (AEA); Bernhard Schlamadinger and Penny Baalman (Climate Strategies); Chris Hope (Judge Business School, Cambridge University); Juan Carlos Castilla-Rubio (Cisco); and Dr Miguel Franco (School of Biological Sciences, University of Plymouth).

We have also greatly benefited from a wealth of expertise provided by academics and researchers, including: Prof Heiko Balzter (University of Leicester); Dr Allan Spessa (University of Reading); Dr Daniel Nepstad (The Woods Hole Research Center); Dr Charles Palmer (Institute for Environmental Decisions); Joanna House (Quest); David Neil Bird (Joanneum Research); Prof Peter Smith (University of Aberdeen); and Prof Shaun Quegan (Centre for Terrestrial Carbon Dynamics, University of Sheffield). The following provided further comments, support and advice throughout the Review: Jonathan Brearley, Stephen Muers and Philippa Benfield (Office of Climate Change); Matthew Owen; Dimitri Zenghelis; Lord Nicholas Stern (London School of Economics and Political Science); Dr Alan Grainger (University of Leeds); and Prof Yadvinder Malhi (University of Oxford).

We are very grateful to the following government departments who have provided invaluable advice and contributions throughout the project: Department for Environment, Food and Rural Affairs; Foreign and Commonwealth Office; Department for International Development; HM Treasury; Department for Business, Environment and Regulatory Reform; the Forestry Commission; British Embassy, Brazil; British High Commission, Guyana; British Embassy, Stockholm; and British Embassy, Paris.

The team have benefited from advice and guidance from a range of organisations and institutions from the public and private sector: ODI, Rights and Resources Initiative, CIFOR, CGIAR, The Nature Conservancy, Rainforest Foundation, the Prince's Rainforest

Project, Global Canopy Programme, Survival International, Forests Philanthropy Action Network, WWF, Down to Earth, The Climate Group, Climate Focus, The Carbon Trust, Fauna and Flora International, Greenpeace, PanEco Foundation, Edinburgh Centre for Carbon Management, Plan Vivo, Wetlands International, Winrock International, The World Conservation Union, Global Witness, Forests and the European Resource Network, World Bank, Syngenta Foundation for Sustainable Agriculture, Conservation International, Energy for Sustainable Development (Camco Group), Forest Resources Management, Forest Peoples Programme, European Space Agency, Deutsche Gesellschaft für Technische Zusammenarbeit, European Space Agency, GAF AG, Food and Agricultural Organisation of the United Nations, FSC, Enviromarkets, EcoSecurities, Eyre Consulting, London Bridge Capital, HSBC, Forest Re, Sustainable Forestry Management Ltd, Carbon Markets and Investors Association, Goldman Sachs, Merrill Lynch, Cisco, TÜV SÜD, CV Starr, Barclays Capital, Morgan Stanley and Carbon Capital.

As part of the evidence gathering exercise, the team visited various rainforest countries and met representatives of many other countries. While it is not possible to mention all those who assisted and advised the team throughout the Review, we would like to make special mention of the following: British Embassy, Brasilia; Brazilian Government ministries including Amazonas State Government's Climate Change Centre; Serviço Florestal Brasileiro; Instituto Nacional de Pesquisas Espaciais (INPE); Instituto Nacional de Pesquisas Amazônia (INPA); organisations based in Brazil including Instituto do Homen e Meio Ambiente da Amazônia (IMAZON), The World Bank, Conselho Nacional dos Seringueiros, Instituto de Pesquisa Ambiental da Amazônia (IPAM), Cikel Brasil Verde Ltda; Reserves and communities including Tapajós-Arapiuns Extractive Reserve, Ilha do Combu, Bélèm, BHC Yaounde; Cameroon Government Ministries including Ministere Des Finances (Minfi), Ministere De Foret Et De La Faune (Minfof), Ministère de l'Environnement et des Forêts; organisations based in Cameroon including WWF, STV, CRTV, Canal 2, *Le jour* Newspaper, *The Post* newspaper, *Mutations* newspaper; Cameroon Reserve Tri-Nationale de la Sangha Reserve; BE Kinshasa, DFID DRC; Minstère de l'Environnement, Conservation de la Nature, Eaux et Forêts (DRC); organisations based in DRC including Centre pour l'Environnement et le Développement, Fédération des Industriels du Bois (FIB), Centre National d'Appui au Développement et à la Participation Populaire (CENADEP), Fauna and Flora International, Reseau Ressources Naturelles, Maniema Libertés, OCEAN, FCS, Dignité Pygmée, USAID CARPE programme, GTZ EITI, l' Université de Kinshasa, Chinese State Forestry Administration and other members of the Chinese study group on FLEGT and forestry (DRC); BE Jakarta, Departemen Kehutanan (Ministry of Forestry, Indonesia), AusAID (Australian Government); and organisations based in Indonesia including CIFOR, World Agroforestry Centre, Carbon Conservation, Birdlife International, Fauna and Flora International and Oxfam. Officials from The European Commission (DG-Environment), Government of Norway (through the International Climate and Forest Initiative) and Government of Sweden (through the Swedish Prime Minister's office and Ministry of Foreign Affairs) kindly provided their views and perspectives.

We would also like to thank Kevin Hogan, Advisor to the President of Guyana, and Kevin Conrad, PNG Special Envoy and Ambassador for the Environment and Climate Change to the United Nations, for their expert insight and advice.

Executive summary

1. The scope, aims and approach of the Review

The Eliasch Review is an independent report to government. It aims to provide a comprehensive analysis of international financing to reduce forest loss and its associated impacts on climate change. It does so with particular reference to the international debate surrounding the potential for a new global climate change deal in Copenhagen at the end of 2009.

The Review focuses particularly on the scale of finance required and on the mechanisms that can, if designed well, lead to effective reductions in forest carbon emissions to help stabilise greenhouse gases in the atmosphere and avoid the worst effects of climate change. It also examines how mechanisms to address forest loss can contribute to poverty reduction, as well as providing incentives to preserve other ecosystem services such as biodiversity and water services.

This Review draws on a large amount of previous research in the literature, responses to a stakeholder consultation exercise and visits to various countries including forest nations in Latin America, Africa and south east Asia. A range of new research and analysis was undertaken by the Review Team and commissioned for the Review from the following international organisations and institutes: AEA; Chatham House; Climate Strategies; CSERGE, University of East Anglia; Ecosecurities; IES; IIASA; IIED; Judge Business School, Cambridge University; LTS International; The Met Office, Hadley Centre; ODI; ProForest; the Royal Botanic Gardens, Kew; School of Biological Sciences, Plymouth University; and the United Nations Environment Programme, UNEP/WCMC.

2. Headline messages

Urgent action to tackle the loss of global forests needs to be a central part of any future international deal on climate change. A deal that provides international forest financing could not only reduce carbon emissions significantly, but also benefit developing countries, support poverty reduction and help preserve biodiversity and other forest services. Forestry, as defined by the IPCC, produces around 17 per cent of global emissions, making it the third largest source of greenhouse gas emissions – larger than the entire global transport sector. In the tropics, it is estimated that an area of forest the size of England is cleared every year, and current annual emissions from deforestation are comparable to the total annual CO_2 emissions of the US or China.

If the international community does nothing to reduce deforestation, modelling for the Eliasch Review estimates that the global economic cost of climate change caused by deforestation could reach $1 trillion a year by 2100. This is additional to the impacts of industrial emissions. Moreover, without tackling forest loss, it is highly unlikely that we could achieve stabilisation of greenhouse gas concentrations in the atmosphere at a level that avoids the worst effects of climate change.

This Review believes that an ambitious international climate change deal should aim to halve deforestation emissions by 2020 and make the forest sector carbon neutral by 2030 – with emissions from forest loss balanced by new forest growth. Reducing deforestation rates significantly will require substantial finance. Nonetheless, even taking this into account, the net benefits of halving deforestation could amount to $3.7 trillion over the long term.

In order to achieve this, a global step change is needed in the way land is used and commodities are produced. Success will rest largely on action at the national level. Demand for agricultural commodities and timber will continue to rise as the world population grows and becomes wealthier. National and international policies will need to shift the way demand for commodities is met away from deforestation and towards more efficient and sustainable methods that ensure forest nations and communities grow and prosper. Improvements in agricultural productivity and the sustainable management of forests will play a key role. Consumer countries can also provide incentives for sustainable production through preferential procurement of sustainably-produced products and increased consumer awareness.

A central element in making this shift work will be the inclusion of the forest sector in global carbon markets. In doing so, the costs of reducing global carbon emissions will be reduced substantially, and lower costs will mean that a more ambitious overall emissions target will be possible. The Review's analysis suggests that including deforestation and degradation (REDD) – and additional action on sustainable management – in a well-designed carbon trading system could provide the finance and incentives to reduce deforestation rates by up to 75 per cent in 2030. With the addition of afforestation, reforestation and restoration (ARR), this would make the forest sector carbon neutral.

In addition, the cost of halving global carbon emissions from 1990 levels could be reduced by up to 50% in 2030 and by up to 40% in 2050 if the forest sector is included in a trading system. This is due to the relatively low cost of forest abatement compared to some mitigation in other sectors. These lower costs could also allow the international community to meet a more ambitious global emissions target.

Full global carbon trading will take time to evolve. Any system should meet the needs of countries at different levels of development, particularly the poorest. In the transition period from 2012, the Review recommends that forestry abatement is supported through a combination of finance from carbon markets and other sources from the public and private sectors.

For this to be successful, four building blocks will be needed:

- **Effective targets dependent on baselines**

  Emissions reductions should be measured against national baselines that provide incentives for action by countries with high historical deforestation rates as well as continued action by those with an effective track record of avoiding deforestation.

- **Robust monitoring and reporting**

    While advances in measuring techniques mean that forest emissions can now be estimated with similar confidence to emissions estimates in other sectors, this will require substantial capacity building in many forest nations.

- **A well-designed mechanism for linking forest abatement to carbon markets; and additional funding from the private and public sector**

    Forest abatement in developing countries needs to be matched with more stringent emissions targets for Annex I countries. Getting this balance right could reduce costs, attain a more ambitious global target, and maintain financial incentives for clean technology transfer to developing countries. The Review shows that, if properly designed, inclusion of the forest sector in the EU ETS should have little or no impact on the EU carbon market price. This would maintain incentives for EU investment in new clean technologies. However, a smooth transition that maintains price stability will mean that additional funding from sources outside carbon markets will be needed in the short to medium term. Under one scenario modelled by the Review, $7 billion could be generated by the carbon markets in 2020 which would leave $11-19 billion to be financed from elsewhere if deforestation were to be halved. Much of this may need to come from international public funding.

- **Strong governance and effective mechanisms for the distribution of finance**

    National governments should take the lead in implementing a successful system to tackle deforestation. Clarifying and securing land tenure user rights, and strengthening institutional capacity at all levels, will be essential. Finance may be directed to national and regional levels, local projects or a combination. The full participation of forest communities will make reforms more likely to succeed and benefit the poor. To help promote transparency, countries may choose to manage carbon revenues through a special fund and should report on the policies and measures they have put in place to reduce the loss of their forests.

In the very short term, developing countries will need substantial support for capacity building to prepare for entry into forest credit schemes. Estimates for this Review suggest that capacity building in 40 forest nations could cost up to $4 billion over five years. This will include three key areas: research, analysis and knowledge sharing; policy and institutional reform; and demonstration activities. If international funding from a combination of carbon markets and other sources is to be effective, the finance will need to be well managed and coordinated. The international community will need to agree on the proportion of finance from different sources. Several funds already exist or are planned, and there is potential for overlap and duplication. The UK should help mobilise international action, working with forest nations, major donors, the UN, World Bank and others to build a coordinated system of multilateral funding. This should build on, and draw together, current multilateral initiatives. Given the risks of climate change, the international community must act swiftly and decisively.

3. Recommendations

Strong and urgent action to tackle forest loss is key to a comprehensive approach to tackling climate change. This Review recommends the following:

Finance

- The international community should aim to support forest nations to **halve deforestation by 2020 and make the global forest sector carbon neutral by 2030**. The international community should provide the necessary finance to meet these goals. A combination of international finance from carbon markets and other sources from the public and private sectors will be needed in the short to medium term.
- As a leading international donor, the UK **should make a significant financial contribution** to tackle global forest loss.
- The **forest sector should be fully included in any post-2012 deal at Copenhagen, with market access provided by emissions trading schemes**. This should be matched by stringent emissions reductions targets for Annex I countries and appropriate supplementarity limits on international credits. A linking mechanism between forest abatement and global carbon trading should be institutionalised as part of a wider global carbon market framework. The international community should agree on the **proportion of finance from different sources**.

Sustainable production

- Forest nations and the international community should undertake **research to better quantify land availability** at global, national and regional scales and determine the most effective country-specific policies for shifting to more efficient, sustainable production of commodities and timber. Policies could include improvements in agricultural productivity in the context of wider sustainability policies, use of idle land and sustainable forest management.
- Consumer countries should examine demand-side policies – for example, through **preferential procurement of sustainably produced products** and increasing consumer awareness, ensuring that this is compatible with WTO rules. This should provide incentives for forest nations to promote sustainable production.

Capacity building

- The international community should support forest nations in **urgent research and analysis** to provide more consistent and accurate data on current emissions from the forest sector.
- Countries with specific expertise in the forest sector should share their knowledge and expertise. In particular, **satellite technology and data management** should be made available to support poorer forest nations in measuring and monitoring changes in forest emissions. This will build capacity for countries to participate in financing mechanisms and provide transparency in reporting emissions reductions.

- Many forest nations will want to undertake policy and institutional reforms in order to create a governance environment in which sustainable land and resource management is possible and profitable. Clarifying and securing land tenure and user rights will be an essential part of this. The international community should provide **urgent support for capacity building** where necessary.
- **Demonstration activities** will be needed to test new approaches and demonstrate how credit mechanisms can be used to make land use more efficient and sustainable, promote REDD and ARR and secure wider social and environmental benefits.
- **International public funds should be coordinated effectively**, avoiding a proliferation of competing mechanisms. The UK and EU should help mobilise international action. The UK Government should work with forest nations, European leaders, major donors, the UN, World Bank and others to build a coordinated system of multilateral funding. This should build on, and draw together, current multilateral initiatives such as FCPF, UN-REDD and FIP.

4.   Chapter summaries

1.   Introduction

Climate change is a major global threat. As carbon emissions rise, so does the likelihood of significant damages to water resources, ecosystems and coasts, as well as the impacts on food supplies and health. To avoid the worst effects of climate change, we should aim to stabilise levels of atmospheric greenhouse gases at 445-490 parts per million CO_2e or less. Achieving this global stabilisation target will require strong and urgent international action on a number of fronts – and forests will need to play a central role.

Forestry, as defined by the IPCC, produces around 17 per cent of global emissions, making it the third largest source of greenhouse gas emissions – larger than the entire global transport sector. Annual forest emissions are comparable to the total annual CO_2 emissions of the US or China. If we do not tackle deforestation, it is highly unlikely that we could achieve a CO_2e stabilisation target that avoids the worst effects of climate change.

Forests also deliver additional ecosystem services such as regulating regional rainfall, flood defense, maintaining soil stability and supporting high levels of biodiversity. Many of these services are crucial for maintaining life and livelihoods, with 1.6 billion people depending on them for their welfare and livelihoods to some extent.

2.   Forests, climate change and the global economy

Forests play an important role in regulating the earth's climate. Deforestation and forest degradation release stored carbon into the atmosphere as CO_2 emissions. The global forest sector produces an estimated 5.8 $GtCO_2$ annually. Deforestation is occurring rapidly in the tropics, where an estimated 13 million hectares – an area the size of England – are converted to other land uses each year. Deforestation in tropical regions generally emits significantly more CO_2 than forests elsewhere in the world.

Modelling for the Eliasch Review estimates that the global economic cost of the climate change impacts of deforestation will rise to around $1 trillion a year by 2100 if unabated. The total damage cost of forest loss for the global economy could be $12 trillion in net present value terms. These costs are additional to climate change damage caused by emissions from other sectors.

3.   The drivers of deforestation

As long as the costs of lost forest carbon and other ecosystem services are not reflected in the price of the products supplied from converted forest land then, in financial terms, forests will often be worth more to landholders cut than standing. Social and institutional conditions operating in many rainforest nations, such as tax breaks and subsidies that encourage deforestation, can exacerbate the economic pressures placed upon forests by demand for timber and agricultural commodities.

The decisions of the developed world, such as whether they purchase non-certified timber and foodstuffs, are just as important a factor for driving deforestation. Biofuels targets could cause additional pressure for forest clearance, unless effective sustainability criteria are applied.

4.   Sustainable production and poverty reduction

A global step change is needed in the way land is used and commodities are produced if forest emissions are to be reduced. Our vision is a sustainable system of global production which can meet increasing demand for commodities and lead to reduced carbon emissions, better livelihoods for the poor and preservation of non-carbon ecosystem services such as biodiversity and water services.

This will require significant policy changes in three main areas. First, at the international level, we need to place a value on forest carbon in a new international deal on climate change. Second, at the national level, governance reforms are required to shift policy incentives towards sustainable production. And third, demand-side policies in consumer countries – for example, through preferential procurement of sustainably produced products and increased consumer awareness – can provide incentives for forest nations to promote sustainable production. The full participation of forest communities and indigenous peoples will make reforms more likely to succeed and benefit the poor.

5.   The costs of mitigation

This Review estimates that the finance required to halve emissions from the forest sector to 2030 could be around $17-33 billion per year if included in global carbon trading. These results are based on various estimates from the literature and from work commissioned by the Review.

Further risk modelling commissioned for the Review provides new evidence of the benefits of taking firm action to reduce forest emissions. Reducing deforestation rates significantly will require substantial finance. Nonetheless, even taking this into account, the net benefits of halving deforestation could amount to $3.7 trillion over the long term (net present value). This is based on the global economic savings from reduced climate change minus the costs involved. The benefits would be even greater if the preservation of other ecosystem services were taken into account.

6. A long-term framework for tackling climate change

There are various mechanisms that could be used to achieve reductions in emissions from the forest sector in the long term, as part of an overall global framework. Of these, a system of cap and trade performs best against the criteria of effectiveness, efficiency and equity. Including reduced emissions from deforestation and degradation (REDD) in a global cap and trade system could reduce deforestation rates by up to 75 per cent in 2030. With the addition of sequestration from afforestation, reforestation and restoration (ARR), this would make the forest sector carbon neutral.

This would have additional benefits for the overall goal of stabilising global emissions. Due to the relatively low cost of forest abatement compared to mitigation in other emitting sectors, the cost of halving global carbon emissions from 1990 levels could be reduced by up to 50 per cent in 2030 and up to 40 per cent in 2050 if the forest sector is included in a global trading system. These lower costs could allow the international community to meet a more ambitious global stabilisation target. Forest carbon finance could also make a significant impact on reducing poverty through increased financial flows to developing countries.

7. The current international climate change framework

The current international climate change framework is a long way from delivering the emissions reductions required for a global stabilisation target necessary to give the world a realistic chance of limiting global warming to 2°C. Further action will be needed from developed and developing countries to meet this goal. Institutional reforms will be needed to include forestry fully into a climate change framework post-2012.

8. Transition to a long term framework

The post-2012 transition path towards a long term goal of global cap and trade will need to meet the needs of sovereign nations at different levels of development, particularly the poorest. The most effective transition path to global cap and trade is likely to be a national, incentive-based approach with increasing finance from emissions trading schemes, but also drawing on additional funding sources while carbon markets grow over time.

In the short term, the main objectives should be capacity building and filling the funding gap. Over the medium term, four building blocks are key: effective national-level targets; robust measuring and monitoring of forest emissions; a well designed system for linking forest credits to carbon markets and other sources of finance; and strong governance. In the long term, the goal should be full inclusion in a global carbon market.

9. Effective targets for reducing forest emissions

The first building block in the transition is an effective system of targets that provide a baseline for issuing credits. A baseline-credit system for non-Annex I countries could initially generate credits for emissions reductions on a no-lose or limited liability basis.

Effective targets for reducing forest emissions need to minimise leakage (a reduction in emissions in one area leading to an increase in emissions in another); ensure real reductions compared to business as usual (additionality); and incentivise action to retain or enhance standing forests.

Baselines should be set at the national level to prevent intra-national leakage. They should take account of a country's historical emissions rate and could also incentivise additional action to protect and enhance forest carbon stocks. This will help ensure that emissions reductions in the global forest sector are additional while acting against international leakage by being inclusive. Baselines should also change over time to help ensure additionality, by means of a renegotiation of baselines linked to an indicative trajectory.

10.   Measuring, monitoring and verifying emissions from forests

The second building block is robust measuring and monitoring of forest emissions reductions. National-level emission inventories need to be comprehensive and internationally consistent to enable verification of emissions reductions. Using appropriate techniques, forest emissions can be estimated with similar confidence to emissions estimates in other sectors. However, this will require substantial capacity building. The Review estimates that $50 million will be needed for a sample of 25 forest nations to set up robust national forest inventories, with a further $7-17 million needed for annual running costs in the following years.

11.   Linking to carbon markets

The third building block for tackling forest emissions in the medium term is a well designed mechanism for linking forest abatement to carbon markets and accessing additional funding from the private and public sectors as carbon finance grows. By finding the right balance in carbon markets between more stringent emissions targets and higher supplementarity limits (the proportion of abatement effort that can be met from non-Annex I country credits), the international community could achieve several key objectives. First, fund significant forest abatement; second, reduce the cost of meeting more stringent global emissions targets; third, provide a strong incentive to invest in new clean energy technologies; and finally support a high level of technology transfer to the developing world.

The EU currently has the largest emissions trading scheme. This Review modelled various scenarios with different reduction targets and supplementarity limits to examine the price impacts of including forest credits. The results suggest that if supplementarity limits are set at 50% or lower in Phase III of the EU ETS, then admitting forest credits into the international credit market should have little or no impact on the EU carbon market price. This is because when restrictions on the use of non-Annex I credits are at this level, more costly EU abatement would still be necessary and would continue to set the price for all units of abatement in the carbon market. More important than the inclusion of forest credits will be the level of supplementarity limit set for international credits in general into the EU market.

This Review also modelled the level of finance that carbon markets could provide for forest abatement in the medium term. One scenario modelled suggests that the global carbon market could supply around $7 billion per year in 2020. This would leave a funding gap of around $11-19 billion in 2020 for halving forest emissions (the range depending on the level of rent received by forest nations), which would need to come from other private and public sources.

During the transition to a comprehensive global cap and trade system, a linking mechanism could perform three important functions: aggregate funding from different sources; manage the risk of reversal of emissions reductions using credits placed in a reserve; and reduce the risk of investing in emissions abatement for forest nations. These functions could be performed by a single institution.

12. Governance and distribution of finance

The fourth building block is strong governance and effective mechanisms for the distribution of finance to reduce forest loss. Sovereign nations need to take the lead in implementing a successful system to tackle deforestation. Key areas of reform include clarifying and securing land tenure rights and strengthening the institutional capacity of national, regional and local institutions. The full participation of forest communities will make reforms more likely to succeed and benefit the poor.

Many policy and programme options exist for reducing emissions from deforestation that do not require cash transfers to individuals. However some options will do so, including transfers to subsistence farmers and foresters. Such transfers will involve costs and capacity requirements which may be challenging for many forest nations in the short term. Capacity building and demonstration activities to test these approaches will be needed.

To help promote transparency, countries may choose to manage carbon revenues through a special fund and should report on the policies and measures they have put in place to reduce deforestation. Premium credits generated from programmes with voluntary higher standards that achieve wider social and environmental goals could be made available for preferential treatment in the market.

13. The funding gap and capacity building

The international community needs to act urgently to tackle climate change and address the global loss of forests. In the short-term, many developing countries will require support for capacity building to prepare for participation in forest market schemes. Estimates for this Review suggest that capacity building in 40 forest nations could cost up to $4 billion over five years. Some countries may be able to self-finance, while others may seek ODA support.

At the same time, a combination of international public and private finance will be needed to meet the medium-term funding gap as carbon markets grow. 'Pump priming' of credit mechanisms will be needed in the short term, using a mix of public and private funds. International public funds for this purpose should be coordinated effectively, avoiding a proliferation of competing mechanisms.

14. Conclusions

Deforestation is progressing rapidly, particularly in the tropics. Firm and urgent action is needed. If not, it is highly unlikely that we can achieve a CO_2e stabilisation target that avoids the worst effects of climate change.

Action on deforestation needs to be taken as part of the international negotiations under the Bali Action Plan towards a global climate change deal in Copenhagen, as well as in the wider context of goals on poverty reduction and the preservation of ecosystem services. A step change is needed in the way land is used and commodities are produced. A shift to more sustainable production will be complex and challenging, but not impossible if the international community acts together effectively.

1. Introduction

Key messages

Climate change is a major global threat. Over the last century, global temperatures have risen by 0.7°C. Sea levels are rising at three millimetres a year and Arctic sea ice is melting at almost three per cent a decade. Continued warming of the atmosphere at the same rate will result in substantial damage to water resources, ecosystems and coastlines, as well as having an impact on food supplies and health.

The economic costs of climate change impacts have been estimated at between 5 and 20 per cent of global GDP and could be considerably higher.

Current evidence suggests that to avoid the worst effects of climate change we should aim to stabilise levels of atmospheric CO_2e at 445-490 parts per million (ppm). Achieving this global stabilisation target will require strong and urgent international action.

The forest sector plays a key role in tackling climate change. Forestry, as defined by the IPCC, accounts for around 17 per cent of global GHG emissions – the third largest source of anthropogenic GHG emissions after energy supply and industrial activity. Forest emissions are comparable to the annual CO_2 emissions of the US or China.

Analysis for this Review estimates that, in the absence of any mitigation efforts, emissions from the forest sector alone will increase atmospheric carbon stock by around 30ppm by 2100. Current atmospheric CO_2e levels stand at 433ppm. Consequently, in order to stabilise atmospheric CO_2e levels at a 445-490ppm target, forests will need to form a central part of any global climate change deal.

In addition to their role in tackling climate change, forests provide many other services. They are home to 350 million people, and over 90 per cent of those living on less than $1 per day depend to some extent on forests for their livelihoods. They provide fuelwood, medicinal plants, forest foods, shelter and many other services for communities.

Forests also provide additional ecosystem services, such as regulating regional rainfall and flood defence and supporting high levels of biodiversity. Maintaining resilient forest ecosystems could contribute not only to reduced emissions, but also to adaptation to future climate change.

The Bali Action Plan provides a roadmap for the negotiation of a new regulatory framework for international action on climate change, following the expiry of the first commitment period of the Kyoto Protocol in 2012. The action plan sets out key areas to be negotiated with a view to reaching a new global climate change deal in Copenhagen at the end of 2009. It recognises the importance of reducing deforestation emissions and a system of international finance to meet this goal.

> The aim of this Review is to examine international financing to reduce forest loss and its associated impacts on climate change. The Review focuses particularly on the scale of finance required and on the mechanisms that can, if designed well, lead to effective reductions in forest carbon emissions to help meet a global stabilisation target. It also examines how mechanisms to address forest loss can contribute to poverty reduction, as well as the importance of preserving other ecosystem services such as biodiversity and water services.

1.1 The impacts of climate change

Climate change is a major global threat. Over the last century, global temperatures have risen by 0.7°C. Sea levels are rising at three millimetres a year and Arctic sea ice is melting at almost three per cent a decade. Continued warming of the atmosphere at the same rate will result in substantial damage to water resources, ecosystems and coastlines, as well as having an impact on food supplies and health.

The economic costs of climate change impacts have been estimated at between 5 and 20 per cent of global GDP, and could be considerably higher.

Warming of the earth's climate system has led to increases in global average air and sea temperatures, rising global average sea levels and widespread melting of snow and ice. As Figure 1.1 illustrates, global average sea level has risen by 3.1 millimetres a year since 1993, with a total 20th century rise estimated at 0.17 metre. Satellite data since 1978 shows that the annual average arctic sea ice extent has shrunk by 2.7 per cent per decade.[1]

A large number of other climatic changes have also been observed, including:

- an increase in the global area affected by drought;
- more frequent heat waves over most land areas;
- increased heavy precipitation events over most land areas;
- an increased incidence of extreme high sea level worldwide.

All the emissions scenarios reported by the Intergovernmental Panel on Climate Change (IPCC) project continued global warming of about 0.2°C per decade for the next two decades. These climatic changes will bring a wide range of impacts, some of which will be irreversible (Figure 1.2).

As global temperatures rise, these impacts will become more severe. Millions of people, particularly the poor, will be exposed to an increased incidence of droughts and floods, food and freshwater shortages, disease and the loss of their livelihoods and homes. Developing countries are particularly exposed to the effects of climate change because their economies are so heavily dependent on climate-vulnerable sectors such as agriculture. In addition to the direct costs to humankind, the IPCC suggests that approximately 20-30 per cent of species assessed so far are likely to be at increased risk of extinction if increases in global average temperature exceed 1.5-2.5°C.[2]

---

1 IPCC (2007) AR4 Synthesis Report
2 relative to 1980-1999

Figure 1.1: Changes in temperature, sea level and northern hemisphere snow cover

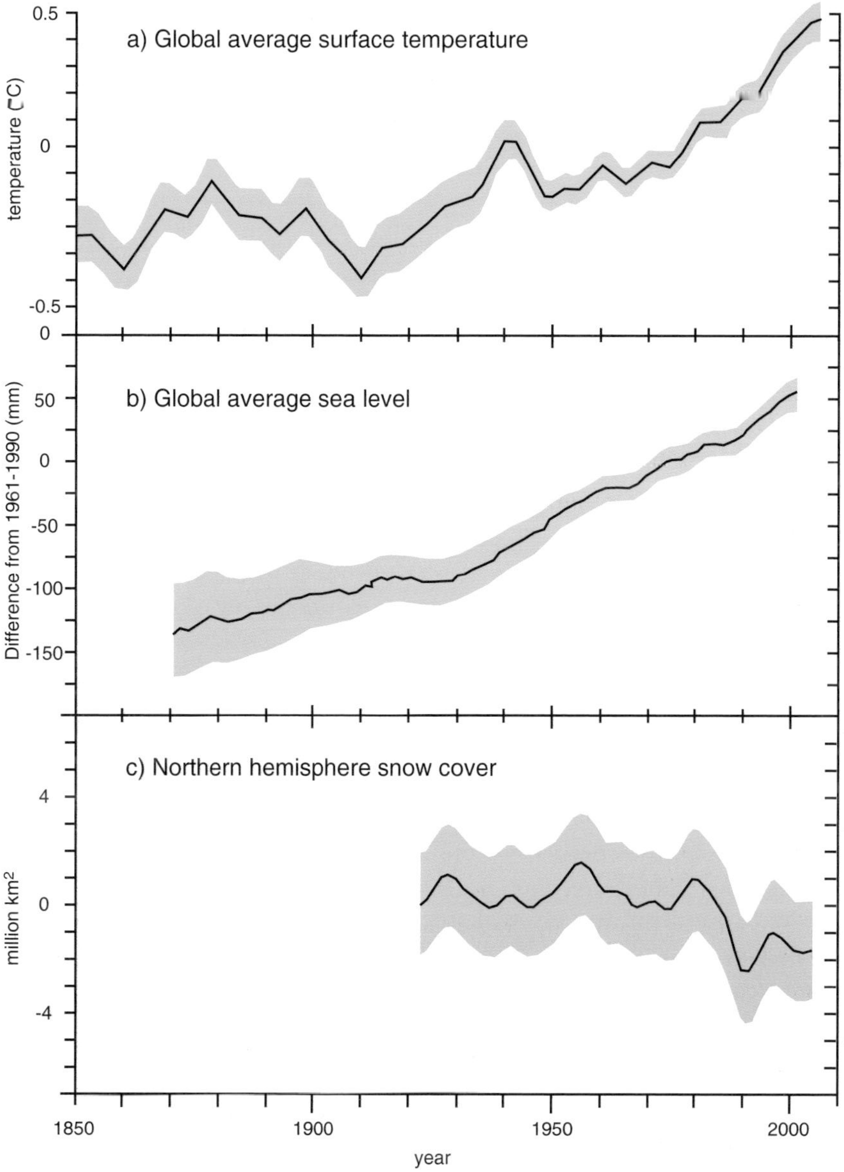

Source: IPCC (2007) AR4 Synthesis Report

Figure 1.2: Examples of impacts associated with global average temperature change

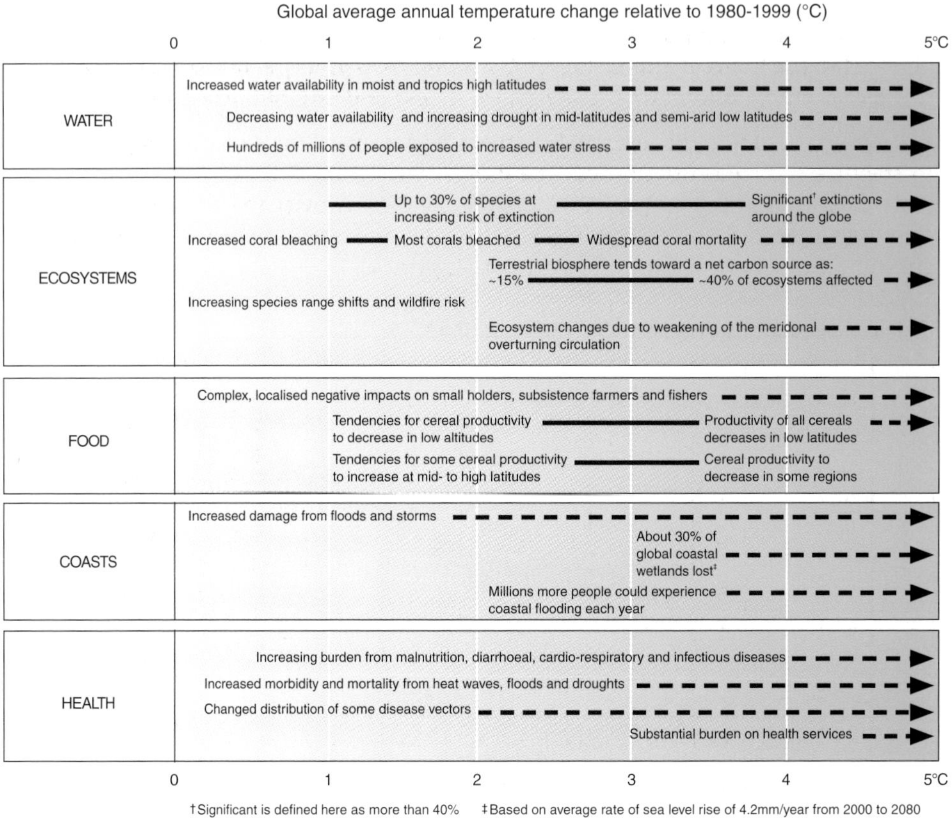

		Global average annual temperature change relative to 1980-1999 (°C)				
	0	1	2	3	4	5°C
WATER		Increased water availability in moist and tropics high latitudes ▶ Decreasing water availability and increasing drought in mid-latitudes and semi-arid low latitudes ▶ Hundreds of millions of people exposed to increased water stress ▶				
ECOSYSTEMS		Increased coral bleaching — Most corals bleached — Widespread coral mortality ▶ Up to 30% of species at increasing risk of extinction — Significant† extinctions around the globe ▶ Terrestrial biosphere tends toward a net carbon source as: ~15% — ~40% of ecosystems affected ▶ Increasing species range shifts and wildfire risk Ecosystem changes due to weakening of the meridonal overturning circulation ▶				
FOOD		Complex, localised negative impacts on small holders, subsistence farmers and fishers ▶ Tendencies for cereal productivity to decrease in low altitudes — Productivity of all cereals decreases in low latitudes ▶ Tendencies for some cereal productivity to increase at mid- to high latitudes — Cereal productivity to decrease in some regions				
COASTS		Increased damage from floods and storms ▶ About 30% of global coastal wetlands lost‡ ▶ Millions more people could experience coastal flooding each year ▶				
HEALTH		Increasing burden from malnutrition, diarrhoeal, cardio-respiratory and infectious diseases ▶ Increased morbidity and mortality from heat waves, floods and droughts ▶ Changed distribution of some disease vectors ▶ Substantial burden on health services ▶				
	0	1	2	3	4	5°C

† Significant is defined here as more than 40% ‡ Based on average rate of sea level rise of 4.2mm/year from 2000 to 2080

Source: IPCC (2007) AR4 Synthesis Report

The Stern Review considered the physical impacts of climate change on the global economy, on human life and on the environment.³ It also estimated the damages of climate change, including integrated assessment models that estimate the economic impacts of climate change.

Stern concluded that business as usual (BAU) climate change will reduce welfare by an amount equivalent to a reduction in consumption per person of between 5 and 20 per cent now and into the future. Subsequent analysis, taking account of the increasing scientific evidence of greater risks, of aversion to the possibilities of catastrophe and of a broader approach to the consequences, suggests the appropriate estimate is likely to be in the upper part of this range.⁴

---

3   Stern (2007)
4   Stern (2008)

1.2 Climate change mitigation

Current evidence suggests that to avoid the worst effects of climate change we should aim to stabilise levels of atmospheric CO_2e at 445-490 parts per million (ppm). Achieving this global stabilisation target will require strong and urgent international action.

Early action could significantly reduce the risk of severe climate change impacts. In order to stabilise the concentration of greenhouse gas (GHG) emissions in the atmosphere, emissions would need to peak and then decline. The lower the stabilisation level, the sooner this peak would have to occur. Table 1.1 summarises the required emission levels and timescales for different stabilisation trajectories.

Table 1.1: Stabilisation scenario characteristics

Category	CO_2 concentration at stabilisation (2005 = 379 ppm)	CO_2-equivalent concentration at stabilisation including GHGs and aerosols (2005 = 375 ppm)	Peaking year for CO_2 emissions	Change in global CO_2 emissions in 2050 (percent of 2000 emissions)	Global average temperature increase above pre-industrial at equilibrium, using 'best estimate' climate sensitivity	Global average sea level rise above pre-industrial at equilibrium from thermal expansion only	Number of assessed scenarios
	ppm	ppm	year	per cent	°C	metres	
I	350–400	445–490	2000–2015	-85 to -50	2.0 – 2.4	0.4 – 1.4	6
II	400–440	490–535	2000–2020	-60 to -30	2.4 – 2.8	0.5 – 1.7	18
III	440–485	535–590	2010–2030	-30 to +5	2.8 – 3.2	0.6 – 1.9	21
IV	485–570	590–710	2020–2060	+10 to +60	3.2 – 4.0	0.6 – 2.4	118
V	570–660	710–855	2050–2080	+25 to +85	4.0 – 4.9	0.8 – 2.9	9
VI	660–790	855–1130	2060–2090	+90 to +140	4.9 – 6.1	1.0 – 3.7	5

Source: IPCC (2007) AR4 Synthesis Report

It is widely suggested that the increase in global temperature that is currently occurring should be stabilised at a maximum of 2°C over pre-industrial levels to minimise the risk of dangerous climate change.[5] The IPCC has indicated in its Fourth Assessment Report that achieving a 2°C target would mean stabilising GHG concentrations in the atmosphere at around 445-490 ppm CO_2 – equivalent (e)[6] or lower.[7] Higher levels would substantially increase the risks of harmful and irreversible climate change (see Figure 1.2).

---

5 eg, EU target for emissions reduction
6 According to IPCC Fourth Assessment Report WG3 (2007) CO_2 – equivalent is 'the concentration of carbon dioxide that would cause the same amount of radiative forcing as a given mixture of carbon dioxide and other greenhouse gases'
7 IPCC (2007) AR4 Synthesis Report

Mitigation efforts over the next two to three decades will to a large extent determine the long-term global mean temperature increase and corresponding climate change impacts that can be avoided. There is widespread agreement that even the most ambitious stabilisation levels could be achieved with the right policies in place and by using a portfolio of technologies that are either currently available or will be available in the coming decades. However, this will require concerted international action across all sectors to:

- reduce demand for emissions-intensive goods and services;
- increase energy efficiency;
- switch to lower-carbon technologies for power, heat and transport;
- take action on non-energy emissions, for example by reducing deforestation.

Different sectors will require different mitigation measures. Energy supply and use and industrial processes will be required to account for 60-80 per cent of GHG reductions, with energy efficiency playing a key role. Reducing current high levels of deforestation and planting new forests will also be needed because forests are a significant contributor to global CO_2 emissions.

The Stern Review estimated that stabilisation at 500-550ppm CO_2e would cost, on average, around 1 per cent of annual global GDP by 2050.[8] This is a significant sum, but is fully consistent with continued growth and development. Even if emissions are stabilised at a lower range of 445-490ppm, the financial benefits of mitigation are likely to be considerable, relative to the costs of unabated climate change.[9] Delay, on the other hand, brings with it a high price. It would require the acceptance of both more intense climate change and, eventually, higher mitigation costs. Weak action over the next 10-15 years would even put stabilisation at 550ppm CO_2e beyond reach – a level associated with significant risks.

1.3 Forests and climate change

The forest sector plays a key role in tackling climate change. Forestry, as defined by the IPCC, accounts for around 17 per cent of global GHG emissions – the third largest source of anthropogenic GHG emissions after energy supply and industrial activity. Forest emissions are comparable to the annual CO_2 emissions of the US or China.

Analysis for this Review estimates that in the absence of any mitigation efforts, emissions from the forest sector alone will increase atmospheric carbon stock by around 30ppm by 2100. Current atmospheric CO_2e levels stand at 433ppm. Consequently, in order to stabilise atmospheric levels at a 445-490ppm target, forests will need to form a central part of any global climate change deal.

The forest sector plays a key role in tackling climate change. Deforestation and forest degradation release stored carbon into the atmosphere as CO_2 emissions. Given the high rates of global forest loss currently, reducing emissions from deforestation and degradation (REDD) would make a major contribution to meeting an emissions stabilisation

---

8   Stern (2007)
9   Hope and Castilla–Rubio (2008)

target. At the same time, afforestation, reforestation and restoration (ARR) increase forest carbon stocks by sequestering and storing carbon from the atmosphere as new forests grow. In addition, natural standing forests maintain carbon stocks and transfers and, under the current climate, act as a carbon sink.

Reducing deforestation and forest degradation is a particular challenge. Forestry, as defined by the IPCC,[10] accounts for around 17 per cent of global GHG emissions (see Figure 1.3). This makes the forest sector the third largest source of anthropogenic GHG emissions after energy supply and industrial activity, and a larger source than the global transport sector. Loss of tropical forest results annually in emissions which are comparable to the total annual CO_2 emissions from the US or China.[11]

Analysis for this Review estimates that emissions from the forest sector alone will increase atmospheric carbon stock by around 30ppm by 2100, whatever efforts are made to mitigate emissions from other sectors.[12] Current atmospheric CO_2e levels stand at 433ppm. Consequently, in order to stabilise atmospheric levels at a 445-490ppm target, forests will need to form a central part of any global climate change deal.

While the forest sector is a major source of CO_2 emissions, reducing forest emissions can be achieved at relatively low cost compared with abatement in other sectors.[13] Furthermore, as the large majority of deforestation occurs in developing countries,[14] any international system that channels finance to reduce deforestation has the potential to help reduce poverty as well as preserve other ecosystem services such as biodiversity and regional rainfall patterns (see following section). This Review examines the design of an international framework for reducing deforestation, focusing particularly on the scale of finance required and the financial mechanisms that can, if designed well, lead to reduced carbon emissions, while providing better livelihoods for the poor and the preservation of non-carbon ecosystem services such as biodiversity and water services.

---

10 IPCC (2007) AR4 Synthesis Report
11 IPCC (2007) WG 1 Chapter 7 reports "the most likely estimate for these [land use change] emissions for the 1990s is 5,800 Mt CO_2/year." The 5.8 Gt CO_2/ year (range: 1.8-9.9 Gt CO_2/year) figure is based on an amalgamation of estimates by Houghton (2003) and Defries et al (2002). Total US CO_2 emissions for 2005 (including LULUCF), as reported to the UNFCCC were 5.3 Gt CO_2. 2005 CO_2 emissions from China (not including LULUCF) were estimated by the IEA (2007) to be 5.1 Gt CO_2.
12 Hope (2008).
13 Stern (2007)
14 FAO (2005)

1. Introduction

Figure 1.3: Sources of global anthropogenic GHG emissions

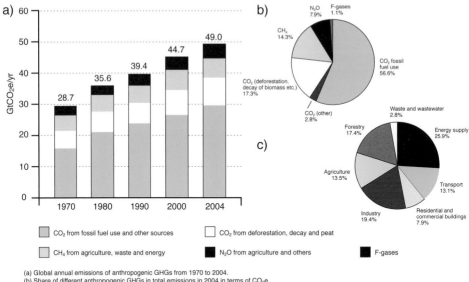

(a) Global annual emissions of anthropogenic GHGs from 1970 to 2004.
(b) Share of different anthropogenic GHGs in total emissions in 2004 in terms of CO_2e.
(c) Share of different sectors in total anthropogenic GHG emissions in 2004 in terms of CO_2e.

Source: IPCC (2007) AR4 Synthesis Report

1.4 Forest communities and ecosystem services

In addition to their role in tackling climate change, forests provide many other services. They are home to 350 million people, and over 90 per cent of those living on less than $1 per day depend on forests to some extent for their livelihood. They provide fuelwood, medicinal plants, forest foods, shelter and many other services for communities.

Forests also provide additional ecosystem services, such as regulating regional rainfall and flood defence and supporting high levels of biodiversity. Maintaining resilient forest ecosystems could contribute not only to reduced emissions, but also to adaptation to future climate change.

Forest ecosystems provide many and varied benefits from their natural resources and processes. The Millennium Ecosystem Assessment[15] categorises forest ecosystem services into five major classes: resources, social services, ecological services, amenities and biospheric services (see Figure 1.4).

---

15 Millennium Ecosystem Assessment (2005)

Figure 1.4: Major classes of forest services

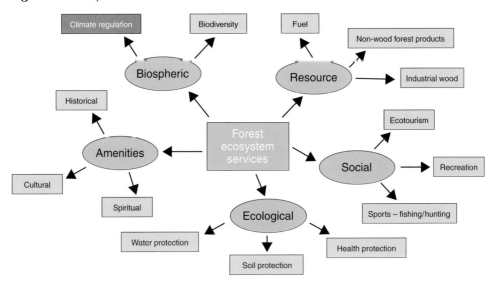

Source: based on Millennium Ecosystem Assessment (2005)

Forests are home to 350 million people around the world and about 60 million indigenous people are almost wholly dependent on forests. More than 1.6 billion people depend to varying degrees on forests for their livelihoods, such as for fuelwood, medicinal plants and forest foods.[16] Wood energy accounts for around 9 per cent of energy consumed worldwide, and up to 80 percent in some developing countries. Bushmeat can account for up to 85 per cent of the protein intake of people living in or near forests.[17] In addition to the use of forests for subsistence food and fuel, community enterprises that generate income from trading sustainably harvested forest resources are on the increase in many countries.

Those who depend on forests are some of the poorest people in the world. Over 90 per cent of those living on less than $1 per day depend on them for their livelihoods.[18] The links between forests and poverty are complex. However, forest communities in remote areas are more likely to be poor, with limited access to services, information and markets. Forest communities are also often politically and economically marginalised, and many lack ownership and use rights over their traditional lands.[19] There is a global trend towards increased recognition of the rights of forest communities and indigenous groups over their land, although progress on human, civil, political and gender rights for forest and indigenous communities is slow.[20]

Global employment in the formal forestry sector is estimated to be around 13 million people,[21] and it has been estimated that for every one job in the formal sector there are another one or two jobs in the informal sector – up to 1 per cent of the global labour

---

16 World Bank (2004)
17 UNDP, UNDESA and World Energy Council (2003)
18 www.fao.org
19 Scherr et al (2003)
20 Chomitz et al (2006)
21 Sunderlin et al (2008)

force.[22] Around 3.5 billion cubic metres of wood are harvested each year from the world's forests and global trade in primary wood products was worth $204 billion in 2006.[23] Forests are also an important supply of other products including latex, handicrafts and medicines.

Leisure time in forests has increased with economic development and urban living. This has led to a growth in social forest services such as recreation, sport and ecotourism.[24] People value forests according to cultural, spiritual and historical factors. These amenities can range from intrinsic and aesthetic value to more geographically specific values relating to the traditional homelands of indigenous people.

Forests provide a range of ecological services including flood protection, pollination, soil formation and erosion control. Forests stabilise their landscapes and offer protection from extreme events such as storms, floods and droughts, which are forecast to become increasingly frequent and intense under future climate change.[25] Resilient forest ecosystems could therefore have an important role in helping people adapt to climate change in the future. They also regulate and supply water and rainfall, which can be particularly important for agriculture. For example, the Amazon forest supplies water to the Rio Plata basin, which generates 70 per cent of the GDP of southern South America through agricultural produce.[26]

Biospheric services include biodiversity as well as climate regulation. Forests contain the majority of terrestrial biodiversity, with tropical forests supporting an estimated 50-90 per cent of the world's species.[27] The value of biodiversity ranges from genetic resources to biological control. Maintaining high levels of biodiversity also aids ecosystem functioning and therefore all other ecosystem services.

1.5 The scope of this Review

The Bali Action Plan provides a roadmap for the negotiation of a new regulatory framework for international action on climate change following the expiry of the first commitment period of the Kyoto Protocol in 2012. The action plan sets out key areas to be negotiated with a view to reaching a new global climate change deal in Copenhagen at the end of 2009. It recognises the importance of reducing deforestation emissions and a system of international finance to meet this goal.

The aim of this Review is to examine international financing to reduce forest loss and its associated impacts on climate change. The Review focuses particularly on the scale of finance required and on the mechanisms that can, if designed well, lead to effective reductions in forest carbon emissions to help meet a global stabilisation target. It also examines how mechanisms to address forest loss can contribute to poverty reduction, as well as the importance of preserving other ecosystem services such as biodiversity and water services.

---

22 Lebedys (2004)
23 www.fao.org
24 Millennium and Ecosystem Assessment (2005)
25 Macqueen and Vermeulen (2006)
26 Mitchell et al (2007)
27 Reid and Miller (1989)

The meeting of the UN Conference of the Parties held in Bali in December 2007 resulted in the adoption of an action plan for formal negotiation of a new global deal on international climate change.[28] The action plan sets out the areas for negotiation and a timetable with a view to reaching a new international agreement on emissions reductions at Copenhagen in December 2009.

The Bali Action Plan represented a major step forward in climate change negotiations, not least by setting out the importance of deforestation as an area of negotiation. The action plan recognises, among other things, the importance of reducing deforestation emissions and technology transfer as mitigation measures as well as support for adaptation (see Figure 1.5). Underpinning these measures will be negotiation of the scale and distribution of international finance. These elements of the action plan are important for providing the incentives for all countries, developed and developing, to participate in a comprehensive emissions reduction system. Without participation of all major emitters, the international community is unlikely to meet the necessary global target for emissions stabilisation. Deforestation is particularly important as many of the poorest countries in the world are affected.

Figure 1.5: Key elements of the Bali Action Plan

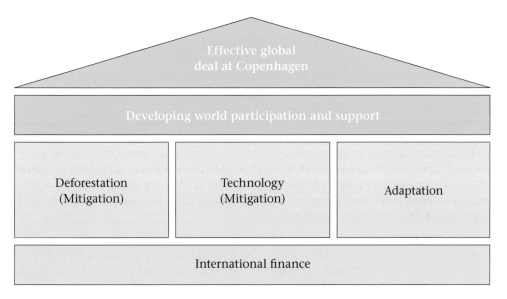

As part of the international debate ahead of Copenhagen, this Review aims to examine international financing to reduce forest loss and its associated impacts on climate change. The Review focuses particularly on the scale of finance required and on the mechanisms that can, if designed well, lead to effective reductions in forest carbon emissions to help meet a global stabilisation target. It also examines how mechanisms to address forest loss can contribute to poverty reduction, as well as the importance of preserving other ecosystem services such as biodiversity and water services.

---

28 Bali Action Plan Decision 1/CP.13 (2007)

1.5.1 The structure of the Review

The first part of this Review begins with an overview of the role of forests in the carbon cycle, the impacts of human activities and the contribution of the forest sector to climate change. It examines business as usual scenarios for carbon emissions from forests and the projected economic costs of deforestation and the resulting climate change. It goes on to explore the economic and policy drivers of deforestation before setting out a vision for sustainable production which reflects the true global value of forests and contributes to poverty reduction. Part I of the Review concludes with an analysis of the costs of moving from current, unsustainable deforestation practices to a more sustainable global system of forest management.

Part II of the Review sets out a long-term international framework for tackling climate change. A variety of systems exist for achieving reductions in deforestation as part of an overall global framework. Many of these systems are potentially valuable tools in tackling climate change and could work in parallel. The two principle options for valuing forest carbon in the long term are taxation and cap and trade. This Review concludes that a global cap and trade system performs best against the criteria of effectiveness, efficiency and equity. The financial flows to forest nations could also have a significant impact on poverty reduction in developing countries. Part II of the Review also looks at the current international framework and explores the advantages and limitations of existing incentives to reduce emissions from forests.

The transition to a long-term goal of global cap and trade will need to be well designed and meet the development needs of countries at different levels of development, particularly the poorest. If the transition path is poorly designed, the long-term goal may not be reached or may be delayed. Four key building blocks should be in place. These include establishing national targets, baselines or reference levels for emissions reductions; measuring and monitoring forest emissions robustly (through national inventories); ensuring that access of forest credits to carbon markets is well designed; and developing a sound system of governance. Part III of the Review discusses these building blocks.

The Review concludes in Part IV by discussing the institutions and systems required to coordinate the transition to a long-term framework and sets out the level of finance required to help build capacity in developing nations and to meet the funding gap which will exist in the short to medium term.

Part I: The challenge of deforestation

Forests play a major role in climate change. Deforestation, forest degradation and activities that increase forest cover all affect climate regulation. This in turn is predicted to have significant impacts on the global economy. Part I of this Review examines the impacts of global deforestation on climate change and the global economy and the challenge of tackling the factors that drive global deforestation.

First, Chapter 2 sets out the importance of forests in regulating the earth's climate. It also explores current rates of change and carbon emissions from forests, and estimates the likely future economic and social costs of continued deforestation.

Chapter 3 then examines the economic, policy and institutional drivers of deforestation, including demand for commodities from deforested land.

Chapter 4 sets out a vision of sustainable production that reflects the true value of forests. It also describes the challenges associated with reducing forest carbon emissions, delivering better livelihoods for forest communities and preserving biodiversity and other ecosystem services.

Finally, Chapter 5 examines the mitigation costs associated with achieving this goal, including the opportunity costs of reduced deforestation for agriculture and timber production. The chapter goes on to estimate the net financial benefits of mitigation compared with the global damages of deforestation.

2. Forests, climate change and the global economy

Key messages

Forests play an important role in regulating the earth's climate through the carbon cycle, storing carbon above and below ground and removing it from the atmosphere as they grow. They currently cover about 30 per cent of the earth's land surface, yet they represent the most significant terrestrial carbon store, containing some 77 per cent of all carbon stored in vegetation and 39 per cent of all carbon stored in soils. Forests sequester and store more carbon per hectare than other types of land cover.

Human activities have significant impacts on the forest carbon cycle. Deforestation and forest degradation release stored carbon into the atmosphere as CO_2 emissions. Afforestation, reforestation and restoration increase forest carbon stocks by sequestering and storing carbon from the atmosphere as new forests grow. Natural standing forests maintain carbon stocks and transfers and, under the current climate, act as a carbon sink.

Since 1980, global forest cover is estimated to have declined by 225 million hectares due to human action. Deforestation is progressing rapidly in the tropics, where an estimated 13 million hectares, an area the size of England, is converted to other land uses each year. By contrast, afforestation and reforestation (A/R) is estimated at 5.5 million hectares every year, mainly in the temperate regions.

The global forest sector produces an estimated 5.8 $GtCO_2$ annually from deforestation, around 96 per cent of which is estimated to come from developing countries in the tropics. Deforestation emits significantly more CO_2 than can be sequestered by an equivalent area of land forested in temperate regions.

Projecting future CO_2 emissions from forests involves a high degree of uncertainty, due to uncertainty over current and future deforestation rates and carbon stocks. A key recommendation of this Review is that greater efforts should be made by the international community to obtain more consistent and accurate data on current emissions from the forest sector.

Modelling commissioned by this Review estimates that the mean damage cost of the climate change impacts of forest emissions will have risen to around $1 trillion a year by 2100. These costs are additional to those caused by emissions from other sectors.

2.1 Forests and the carbon cycle

Forests play an important role in regulating the earth's climate through the carbon cycle, storing carbon above and below ground and removing it from the atmosphere as they grow. They currently cover about 30 per cent of the earth's land surface, yet they represent the most significant terrestrial carbon store, containing some 77 per cent of all carbon stored in vegetation and 39 per cent of all carbon stored in soils. Forests sequester and store more carbon per hectare than other types of land cover.

Carbon is continuously cycled between oceanic, terrestrial and atmospheric reservoirs. The carbon cycle involves stocks, where the carbon remains for a period of time, and fluxes, which transport carbon between stocks (see Figure 2.1). Processes in the cycle operate over many timescales: sedimentation, for example, can act over millions of years, while respiration takes a matter of seconds. Living organisms, soils and detritus (dead organic matter) play an important part in the cycle by removing carbon from and returning it to the atmosphere in the form of CO_2, as well as acting as a carbon store.

Figure 2.1: The natural carbon cycle

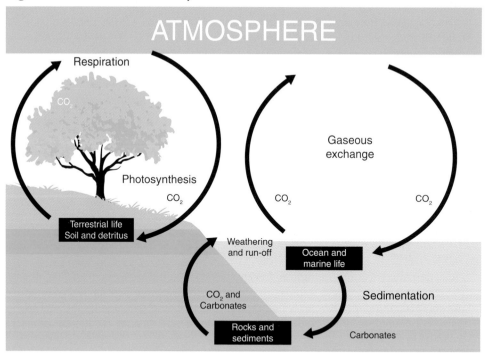

As forests grow and increase their biomass, they absorb carbon from the atmosphere and store it in plant tissue. This process is known as carbon sequestration. Despite frequent exchanges of carbon between forest biomass, soils and the atmosphere, a large amount is always present in leaves and woody tissue, roots and organic matter in soil. This quantity of carbon is known as the carbon store. Established forests are currently acting as a carbon sink because they are absorbing more carbon through photosynthesis than they release through respiration.

Forests currently cover about 30 per cent of the earth's land surface, yet they represent the most significant terrestrial carbon store, containing an estimated 77 per cent of all carbon stored in vegetation and 39 per cent of all carbon stored in soils.[1] They also store twice as much carbon than is present in the atmosphere.[2] The amount of carbon in a forest varies geographically but all forests store, on average, more carbon per hectare than other types of land cover (see Box 2.1, Figure A). Tropical rainforest can contain at least four times more carbon per hectare than cropland in the tropics.[3] When growing, forests generally have higher carbon sequestration rates than other types of vegetation (see Box 2.1, Figure B).[4]

> **Box 2.1: Carbon stocks and net primary productivity in forests and other types of land cover**
>
> Forests have higher carbon stocks per hectare than other types of land cover (see Figure A). Total average forest carbon stocks vary geographically between tropical, temperate and boreal regions, as does the relative proportion of carbon held in vegetation and soil.
>
> **Figure A:** Average carbon stocks for different types of land cover
> *Note: averages taken using two datasets and error bars represent the maximum values*
>
>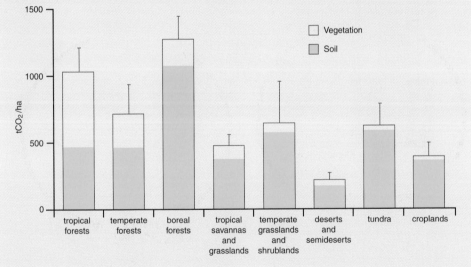
>
> *Source: data from IPCC (2001) WG1, Chapter 3*

---

1   WBGU (1998), IPCC (2007) WG3, Chapter 9
2   WBGU (1998), IPCC (2007) WG1, Chapter 7
3   Houghton (1999)
4   Houghton (2007); IPCC (2006)

Through the process of photosynthesis, plants convert CO_2 into biomass using the energy from sunlight. Net primary productivity is the net flux of carbon from the atmosphere into green plants per unit time. It is, effectively, a measure of carbon sequestration. Forests store more carbon per hectare than other types of land cover, because lignified tissues, ie wood, decompose more slowly than soft tissues.[5] In addition, tropical regions have higher sequestration rates than other regions largely because of the longer growing season. Figure B shows average net primary productivity for different types of land cover.

Figure B: Average net primary productivity (sequestration) for different types of land cover

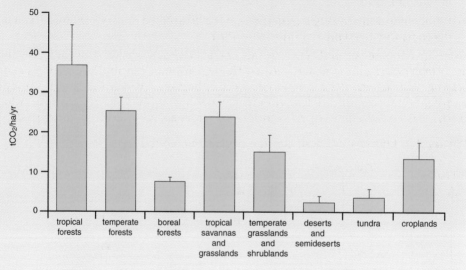

Note: averages taken using two datasets and error bars represent the maximum values
Source: data from IPCC (2001) WG1, Chapter 3

2.2 Impacts of human activities on the forest carbon cycle

Human activities have significant impacts on the forest carbon cycle. Deforestation and forest degradation release stored carbon into the atmosphere as CO_2 emissions. Afforestation, reforestation and restoration increase forest carbon stocks by sequestering and storing carbon from the atmosphere as new forests grow. Natural standing forests maintain carbon stocks and transfers and, under the current climate, act as a carbon sink.

---

5 Franco (2008)

2.2.1 Deforestation and degradation

Both deforestation and forest degradation reduce the amount of forest cover and vegetation and release substantial emissions of CO_2 to the atmosphere.[6] The effects on the carbon cycle can be summarised as follows (see Figure 2.2):

a) Carbon stored in living and dead plant material is released as CO_2 by burning or decomposition.

b) Carbon is released from the oxidation of the soil.

c) Sequestration of CO_2 from the atmosphere is reduced.

d) The transfer of carbon from vegetation to litter, deadwood and soil is reduced. Carbon stored in forest soils is often equal to, or greater than, that stored in above ground biomass.

e) Carbon is lost in the longer term through the breakdown of harvested wood, at a rate dependent on the nature of the end product.

Other greenhouse gas emissions can also be associated with forest disturbance, such as nitrous oxides and methane released through biomass burning.[7] Further emissions leading on from the deforestation depend on the use to which the land is put after clearing. If it is converted to farmland then additional indirect emissions might include methane from cattle farming and nitrous oxide from applying fertiliser to cropland.[8]

Figure 2.2: Effects of deforestation and degradation on the carbon cycle

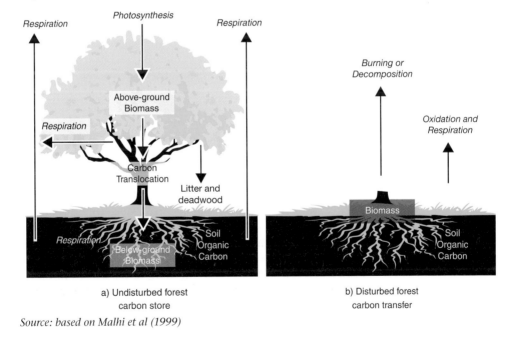

a) Undisturbed forest carbon store

b) Disturbed forest carbon transfer

Source: based on Malhi et al (1999)

---

[6] We use the UNFCCC definition of deforestation as the direct human-induced conversion of forested land to non-forested land. Degradation occurs when a forest is damaged – for example by cutting down a proportion of the trees, which reduces forest carbon stocks and sequestration.

[7] van Amstel and Swart (1994)

[8] Houghton in Moutinho and Schwartzman (2005)

A range of human activities can lead to deforestation and degradation and these can affect the carbon cycle in different ways and to different extents.

Slash and burn is used to clear forest for farming and involves cutting down trees and burning the under-storey, leaving land clear for cultivation or pasture. Using this method, 90-100 per cent of carbon stored in the vegetation is released immediately, as well as 25 per cent of soil carbon to a depth of one metre. This release of soil carbon is significant, particularly in the highly carbon-rich soils of tropical peat forests.[9] This land use change can last decades or it can be shorter term if land is cultivated for a few years and then abandoned once soil fertility declines. Abandonment allows the re-growth of vegetation: primary forest is replaced by either rough pasture or secondary forest, both of which have lower carbon stocks and are often re-cleared at rates that rival primary forest deforestation.[10]

Logging involves the removal of trees for land management or commercial activities. Clear-cut logging removes large areas of trees, leaving behind stumps and litter material. This practice can release 40-60 per cent of the total carbon stored, the majority coming from the vegetation.[11] There is, in addition, a slower release of carbon associated with the decay of dead plant material left on-site. In the long term, carbon is also lost through the breakdown of the harvested wood, the rate of this depending on the end product.

Selective logging – where individual valuable trees are removed – can result in degradation rather than deforestation. In most tropical forests (with the exception of some Asian forests) logging is generally selective because most trees do not have a high commercial value. In principle, selectively logged forests allowed to recover can still have a high biodiversity value and provide many ecosystem services. But in practice, selective logging often leads to complete forest clearance.

Deforestation and degradation also have indirect effects on carbon emissions. Following logging, forests have increased sensitivity to burning:[12] reduced canopy cover allows increased light penetration which dries the organic debris in the under-storey, increasing its flammability. Removing forest canopy also reduces the amount of precipitation that is intercepted and evaporated. This can increase run-off, causing soil erosion and damage to the remaining vegetation.[13] Forests also become fragmented by clear-cutting, selective logging and the associated access roads. Trees on forest edges are more likely to suffer water stress leading to dieback and enhanced fragmentation.[14]

2.2.2 Afforestation, reforestation and restoration (ARR)

Afforestation is the planting of new forests on lands that, historically, have not contained trees. Reforestation describes the establishment of trees on land that has been cleared of forest within the recent past.[15] Restoration is the enhancement of damaged forest to re-establish a forest to its natural structure and carbon stock.[16] This is generally achieved

---

9   Joosten and Couwenberg (2007)
10 Hirsch et al (2004); Steininger (2004)
11 Sajwaj et al (2008)
12 Nepstad et al (1999)
13 Nepstad et al (1994)
14 Giambelluca et al (2003)
15 As defined by IPCC (2000)
16 UNEP-WCMC FRIS (2008)

through planting, seeding or assisting natural regeneration of the structure, productivity and species diversity of the forest originally present. ARR activities sequester carbon from the atmosphere and increase forest carbon stocks.

In recent decades, the area of forest in mid-latitudes has expanded due, in part, to an increase in afforestation projects. China, for example, has been implementing an afforestation scheme in order to reduce land degradation and provide timber.[17] Despite this expansion, forest plantations still account for less than 5 per cent of the total forest area,[18] and current estimates suggest that, as yet, ARR activities have not had a significant impact on the global terrestrial carbon sink (see Box 2.2).[19]

Box 2.2: The carbon sink

Standing forests, particularly in the tropics, act as a natural carbon sink, absorbing more carbon from the atmosphere through photosynthesis than they release through respiration.

Carbon sinks play a significant role in offsetting some of the total anthropogenic emissions of CO_2. The IPCC shows that the current annual rise in CO_2 concentration in the atmosphere is only about 57 per cent as high as it would have been from fossil fuel emissions without the sink effect (see Figure A) and only about 40 per cent as high as it would have been from all emissions including those from land-use change.[20] This suggests that without carbon absorption by forests and other carbon sinks (in particular the oceans), the rise in CO_2 caused by anthropogenic emissions would have been considerably higher.

One simple climate-carbon cycle model simulating historical emissions has suggested that if the tropical forest carbon sink had not been present, the CO_2 increase due to past emissions would have been 10 per cent higher.[21]

Old-growth tropical forests are estimated to absorb about 4.4 +/- 1.5 $GtCO_2$ a year through the sink effect, equivalent to 15 per cent of annual anthropogenic greenhouse gas emissions.[22] Deforestation leads directly to emissions of CO_2 to the atmosphere but it also results in the loss of this forest carbon sink. This could have an additional effect on the rate of rise of CO_2, termed the 'land use amplifier',[23] since it amplifies the effect of emissions from other sources.

The removal of a forest carbon sink has long-term implications for the atmospheric CO_2 concentration. While the emissions resulting from the removal of an area of forest occur over a short period, the absence of the sink persists unless the forest is replaced. For example, if a hectare of forest was sequestering three tonnes of CO_2 per year,[24] the

---

17 Li (2004)
18 FAO (2006)
19 IPCC (2007) WG 1 Chapter 7
20 Betts et al (2008)
21 Betts et al (2008) using a model by Huntingford et al (2008)
22 IPCC (2007) WG 1 Chapter 7
23 Gitz and Ciais (2003)
24 Phillips et al (1998)

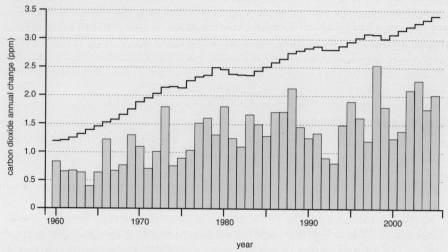

Figure A: Comparison of annual fossil fuel CO_2 emissions with annual rise in atmospheric CO_2 concentration

Note: stepped line shows annual fossil fuel CO_2 emissions. Bars show the annual rise in atmospheric CO_2 concentration.

Source: IPCC (2007) WG1, Chapter 7

presence of this forest over 40 years would result in the absorption of 120 tonnes of CO_2. If, in addition, that hectare of forest stored carbon equivalent to 600 tonnes of CO_2,[25] deforestation of that hectare would not only result in the emission of most of the 600 tonnes of CO_2 in the short term, but would also mean that the forest would not absorb the 120 tonnes of CO_2 over a 40 year period. The long-term impact on cumulative net emissions over those 40 years is therefore 20 per cent greater than would have been expected if only the initial emissions were taken into account.

---

25 Using an average estimate of plant carbon density from House et al (2002)

2.3 Impacts of forests on climate change

Since 1980, global forest cover is estimated to have declined by 225 million hectares due to human action. Deforestation is progressing rapidly in the tropics, where an estimated 13 million hectares, an area the size of England, is converted to other land uses each year. By contrast, afforestation and reforestation (A/R) is estimated at 5.5 million hectares every year, mainly in the temperate regions.

2.3.1 Rates of deforestation, degradation and ARR

Until the mid-20th century, most deforestation occurred in temperate regions. However in recent decades, abandonment of agricultural land in Western Europe and North America has led to some natural reforestation. Conversely, deforestation is now progressing rapidly in the tropics. Since 1980, global forest cover is estimated to have declined by 225 million hectares due to human action.[26] In the tropics, an estimated 13 million hectares, an area the size of England, are converted to other land uses each year.[27]

Although the overall area under forest has declined, the rate of forest loss has actually slowed (see Table 2.1 below): current net deforestation rates are 18 per cent lower than in the 1990s and 26 per cent lower than in the 1980s. The lower rate of global net forest loss most probably reflects increased forest cover in the mid-latitudes rather than a decrease in deforestation in the tropics. For example, the table below shows a net increase in forest area in Asia between 2000 and 2005, mainly due to afforestation projects in China. Although afforestation has been extensive in higher latitudes, these forests tend to sequester significantly less CO_2 than tropical deforestation emits (see following sections).

Table 2.1: Estimated annual change in forest area since the 1980s by continent and globally

Region	Annual change in forest area (10,000 km²)		
	1980s	1990s	2000-2005
Africa	- 28	- 44	- 40
Asia	- 9	- 8	10
Europe	2	9	7
North and Central America	- 12	3	- 3
Oceania	- 0.4	- 4	- 4
South America	- 52	- 38	- 43
Global	- 99	- 89	- 73

Source: FAO (1990); IPCC (2007) WG 3, Chapter 9, after FAO (2006)

---

26 FAO (1990); FAO (2006). This figure is net deforestation: it includes deforestation and forestation activities.
27 FAO (2006). This is the average annual area of all countries that have reported net deforestation over 2000 to 2005

Forest degradation is harder to detect than deforestation, and there are few studies looking at degradation at a global level. However, it is clear that degradation is widespread, especially in the tropics in areas adjacent to deforestation. When an area is deforested for logging, agriculture or road-building, it can result in as much as twice the surrounding area being degraded due to forest fragmentation and the extension of human activities into relatively undisturbed areas.[28]

While the overall global area under forest has declined, rates of ARR have increased in the mid-latitudes. The Food and Agriculture Organisation (FAO) estimates that 5.5 million hectares of land is being forested every year, mainly through the expansion of plantations.[29] In addition, agricultural intensification in mid to high latitudes has reduced the area of land used for agriculture and this abandoned land has been subject to natural forestation – although of course more intensive agricultural production has associated carbon costs, which to some extent offset the CO_2 sequestered by this natural forestation.

While forests have a role in regulating climate, changes in local climate can in turn lead to changes in forest characteristics. The effects of global warming on local climate and the responses of ecosystems to these changes are uncertain. However, studies point to long-term climate impacts such as increased Amazonian drying and forest die-back.[30]

2.3.2 CO_2 emissions and sequestration

The global forest sector produces an estimated 5.8 $GtCO_2$ annually from deforestation, around 96 per cent of which is estimated to come from developing countries in the tropics. Deforestation emits significantly more CO_2 than can be sequestered by an equivalent area of land forested in temperate regions.

Annual emissions from deforestation stand at 5.8 $GtCO_2$,[31] and around 17 per cent of global anthropogenic greenhouse gas emissions come from the forestry sector as a whole, including emissions from biomass decay, drained peat and peat fires. The forest sector is the third largest source of emissions after energy supply and industry.[32] On a global scale, forestry is a larger emitter than the transport sector.

Table 2.2 below shows net CO_2 emissions (emissions from deforestation minus sequestration from A/R) from the forest sector in the 1980s and 1990s. The range of estimates given is high because there are uncertainties associated with estimating deforestation rates and forest carbon stocks (see Betts et al 2008). The figures are also likely to underestimate emissions since they do not include those from forest degradation, which can also be significant. As well as affecting the carbon cycle, changes in forest cover can also affect the climate in other ways, through changes to surface albedo, evaporation rates and aerosol emissions (see Box 2.3).

Around 96 per cent of emissions from deforestation are in developing countries in the tropics, while over 60 per cent of sequestration from forestation is in the temperate and boreal regions.[33] Both tropical deforestation emissions and sequestration from non-tropical regions have increased since the 1980s. However, temperate forests do not sequester

---

28 Peres et al (2006)
29 FAO (2006)
30 Malhi et al (2008)
31 IPCC (2007) WG 3 Chapter 9
32 IPCC (2007) AR4 Synthesis Report
33 Houghton (2003)

as much CO_2 as is released in emissions through tropical deforestation. And the amount of CO_2 sequestered each year is only a fraction of the amount released each year through deforestation and degradation.

Tropical deforestation not only produces substantial carbon emissions but it also reduces the size of the forest carbon sink (see Box 2.2). This can lead to an 'amplifying' effect on the rate of increase in the global atmospheric CO_2 stock.

Table 2.2: CO_2 emissions from land use change in the 1980s and 1990s ($GtCO_2$ per year)

	Tropical Americas	Tropical Africa	Tropical Asia	Pan-Tropical	Non-tropics	Total globe
1980s AR4	2.2 (1.1-2.9)	0.7 (0.4-1.1)	2.2 (1.1-3.3)	4.8 (3.3-6.6)	0.2 (-1.5-2.2)	5.1 (1.5-8.4)
1990s AR4	2.6 (1.5-3.3)	1.1 (0.7-1.5)	2.9 (1.5-4.0)	5.9 (3.7-8.1)	-0.1 (-1.8-1.8)	5.9 (1.8-9.9)

Note: Positive values indicate net emissions and negative values net sequestration. Numbers in parentheses are ranges of uncertainty.
Source: Table adapted from IPCC (2007) and converted from GtC per year to $GtCO_2$ per year. The figures are IPCC AR4 best estimates calculated from the mean of Houghton (2003) and Defries et al (2002), the only two studies covering both the 1980s and the 1990s. For non-tropical regions where Defries et al has no estimate, Houghton was used.[34]

> **Box 2.3: Additional impacts of forests on climate change**
>
> As well as affecting the carbon cycle, changes in forest cover can also modify the terrestrial surface energy balance, including via the albedo, evaporation rates and aerosol emissions, all of which can exert local warming or cooling effects. (See Betts et al (2008) for more details).
>
> - *Albedo* is the proportion of solar radiation reflected back into space, commonly expressed as a percentage. A low albedo means that very little solar radiation is reflected while more is absorbed by the surface, which acts to warm the land. A forested landscape generally has a low albedo of around 10-20 per cent, while grasslands and croplands have higher albedos of around 40-50 per cent. A high albedo means that a high proportion of solar radiation is reflected back into space while less is absorbed, so the surface receives less warming from the sun. Snowy landscapes can have an albedo of 80 per cent or more because snow is highly reflective.[35]
> - Water on the surface of land or vegetation tends to evaporate back to the atmosphere, and this *evaporation* exerts a cooling influence on the land surface, as well as providing moisture to the atmosphere and enhancing cloud formation. Evapotranspiration also takes place when plants draw up moisture from the soil, which then evaporates into the atmosphere through microscopic pores in the leaf surface. Forests have higher rates of evapotranspiration than other types of land cover and evapotranspiration rates are also higher in warmer regions.

---

34 IPCC (2007) WG 3 Chapter 9
35 Harding and Pomeroy (1996)

> - The burning of biomass (for example through slash and burn deforestation) produces soot *aerosol emissions*. Deforestation in dry regions can also produce dust aerosol emissions. Both soot and dust absorb solar radiation, leading to a warming effect. Although the effects are complex and have not yet been investigated in depth,[36] deforestation through slash and burn and deforestation in regions which are drying due to climate change may exert additional warming effects.
>
> The interactions involved are complex, and the relative importance of each of these biophysical processes depends on local conditions and can vary with season and location. In cold regions, evaporation rates are lower and albedo is the dominant factor. The presence of forests in cold regions with substantial winter snow can exert an overall warming effect on the local climate (forest cover having a lower albedo than snow) while deforestation in mid-latitudes has an overall cooling effect because of increases in surface albedo. In tropical regions, although albedo is a significant factor, evaporation is more significant. Deforestation in the tropics exerts an overall warming effect through reduced evaporation rates. This warming effect is additional to that caused by increasing greenhouse gas concentrations.
>
> Because there is still much uncertainty in this area, the climatic impacts of albedo, evapotranspiration and atmospheric aerosols are not included in the modelling used in this Review unless otherwise stated.

2.4 Modelling future impacts

2.4.1 Business as usual scenarios

Projecting future CO_2 emissions from forests involves a high degree of uncertainty, due to uncertainty over current and future deforestation and degradation rates and carbon stocks. A key recommendation of this Review is that greater efforts should be made by the international community to obtain more consistent and accurate data on current emissions from the forest sector.

Projecting rates of deforestation, degradation and ARR under a business as usual (BAU) scenario and calculating the resultant estimated CO_2 emissions is needed to understand the projected effects of reduced forest loss. There are many models of the forest sector that are capable of forecasting future emissions. Figure 2.3 shows two very different examples of business as usual forecasts of forest sector emissions. Although different, both start from the 1.8-9.9 range used by the IPCC and they both show a decline in emissions for the next four decades. Further, their cumulative emissions from deforestation to the end of the 21st century are broadly similar.

Houghton's[37] projections estimate forest emissions at around 8 $GtCO_2$ in 2000, declining over time to around 2 $GtCO_2$ in 2100. In contrast, the IPCC SRES A2 projections[38] estimate emissions at around 4 $GtCO_2$ in 2000, with emissions increasing from

---

36 Woodward et al (2005)
37 Houghton (2003)
38 Betts et al (2008) used the IMAGE model by Strengers et al (2004) using the SRES A2 BAU scenario laid out in Nakicenovic et al (2000) and carbon stock assumptions from House et al (2002)

around 2050, reaching around 10 GtCO$_2$ in 2100. The variation in the forecasts indicates the level of uncertainty associated with current emissions from deforestation, with many estimates lying between these two forecasts. As well as reflecting the uncertainties in current emissions, the two graphs also indicate the large uncertainties over the changing rate of forest emissions in the future. This is partly a consequence of different definitions of business as usual and partly a result of underlying differences in the assumptions and modelling approaches used.

Figure 2.3: Different forest emissions projections from Houghton and SRES A2

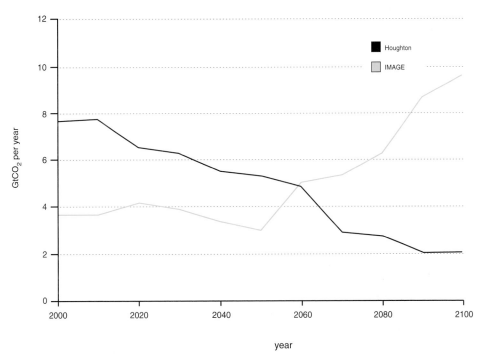

The IPCC has attempted to deal with this uncertainty by modelling emissions under a range of different scenarios known as SRES scenarios. The IMAGE[39] model has been used to project future forest cover changes under different SRES scenarios. Each SRES scenario reflects different assumptions about GDP growth, population growth and about changes in technology. Under four commonly used SRES scenarios, IMAGE projected some areas of deforestation in all tropical regions, and also some areas of A/R in all regions. Overall, tropical forest cover in Africa and Asia is projected to decrease continuously to 2050 in all scenarios. In Latin America, total tropical forest cover in Amazonia decreases by 2050 in only one scenario. In the other three, deforestation is offset by A/R at the continental scale, leading to overall forest gain by 2020 and 2050. The net global land cover changes in these four SRES emissions scenarios are projected to increase atmospheric CO$_2$ concentrations by between 20-127ppm by the end of the 21st century.[40]

---

39 Nakicenovic et al (2000). SRES includes a large number of scenarios. Here we focus on four scenarios knows as A1B, A2, B1 and B2. A number of integrated assessment models were used in SRES, but here we focus on land use from IMAGE (Strengers et al, 2004).
40 Sitch et al (2005)

Given the uncertainty over deforestation projections, the models in the next section, which project the contribution of deforestation to the impacts of climate change, use the Houghton projections in the core scenario, and then the SRES A2 projections to cross-check the results. As an extra level of testing, the models used are probabilistic to give a range of outcomes with confidence intervals.

2.4.2 Costing the climate change impacts of deforestation

Modelling commissioned by this Review estimates that the mean damage cost of the climate change impacts of forest emissions will have risen to around $1 trillion a year by 2100. These costs are additional to those caused by emissions from other sectors.

While several estimates of the mitigation costs of reducing deforestation exist (see Chapter 5), no estimates of the damage costs of the impacts of forest emissions had been published at the time of writing. The Eliasch Review therefore commissioned the modelling of these damage costs using the PAGE model.[41] This probabilistic integrated assessment model was used by the Stern Review to estimate the value of climate change impacts from 2000 to 2200. Figure 2.4 below summarises the steps involved in integrated assessment modelling, although most models do not represent all the steps. The PAGE model does not represent direct impacts; instead, impacts (economic, social and environmental) are evaluated in terms of an annual percentage loss of GDP in each region.

Figure 2.4: Flow chart of steps involved in integrated assessment models of climate change

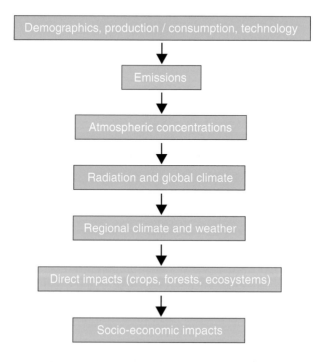

Source: Parson and Fisher-Vanden (1997)

---

41 Hope (2006)

The BAU emissions data used in the core model run is from Houghton[42] for deforestation emissions and the IPCC SRES A2 non-intervention scenario[43] for emissions from other sectors. The full details of the modelling and results are set out in a background paper to this Review.[44]

The modelling shows that with deforestation, CO_2 concentrations are on average around 30ppm higher in 2100 and throughout the 22nd century (see Figure 2.5 below). The IPCC[45] estimates that stabilising the CO_2e concentration within the 445-490 range would result in a global average temperature increase of 2.0-2.4°C. The current CO_2e concentation is around 433ppm.[46] If we are to stabilise levels at around 445-490ppm CO_2e, the international community will need to limit any further global increases to 12-57ppm. Consequently, a contribution of 30ppm from the forest sector is highly significant.

Figure 2.5: Increase in CO_2e concentration by date as a consequence of BAU deforestation

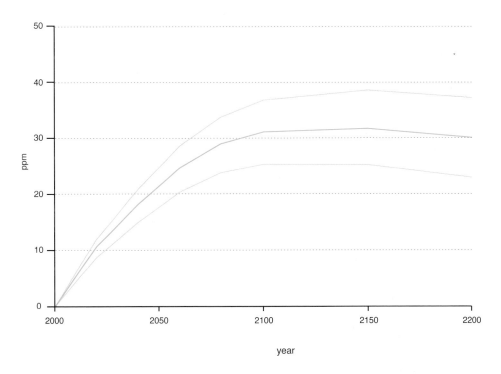

The thick line is the mean result, and the thinner lines are the 5 and 95% points on the probability distribution.

Source: Hope (2008)

---

42 Houghton (2003)
43 Nakicenovic et al (2000)
44 Hope (2008)
45 IPCC (2007) AR4 Synthesis Report
46 EEA (2008) reports that the concentration was 433ppm in 2006.

The increased temperature resulting from deforestation leads to annual damages from global impacts that are, on average, around $1 trillion higher in 2100 (see Figure 2.6). For comparison, global world product was about $45 trillion in 2000, and will rise to $340 trillion in 2100 in the SRES A2 scenario.

Figure 2.6: Increase in annual global impacts by date as a consequence of BAU deforestation

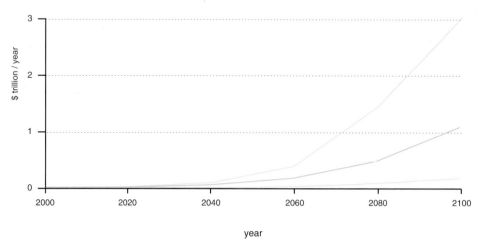

Source: Hope (2008)

In order to represent the sum of all the future damage costs of deforestation in today's terms, the net present value (NPV) of the costs was calculated. This calculation uses a discount rate to give less weight to future damage costs than damage costs in the present. With this discounting,[47] the mean NPV of the climate change impacts of deforestation from 2000 to 2200 is around $12 trillion (see Figure 2.7a). The vertical lines in this and the other graphs in Figure 2.7 represent the 5 per cent and 95 per cent points on the probability distribution.

Although very significant, it is important to bear in mind that the loss of forest carbon services from deforestation is just one lost ecosystem service among many. Other lost services include water regulation and biodiversity (see Chapter 1). It has been estimated that the damage cost of all forest ecosystem services lost in just one year currently amounts to €1.35–3.1 trillion.[48]

---

47 A pure time preference rate of 0-2% was used (most likely value of 1) with an equity weight of 0.5-2 (most likely value of 1). Using these values gives discount rates for the various regions modelled of 1.9-3.6%.
48 Braat and Ten Brink (2008)

Figure 2.7: Forest emissions, climate change damage costs

(a) Damage costs of the climate change impacts from BAU forest emissions (using Houghton and A2 BAU emissions projections)

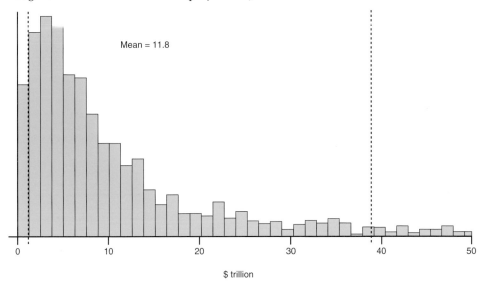

(b) Climate change damage costs of deforestation (using IPCC SRES A2 BAU emissions projections for forest and other sectors)

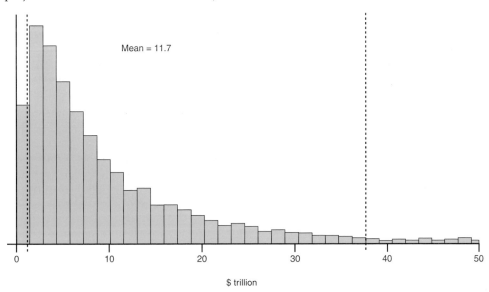

(c) Climate change damage costs of deforestation (using Houghton and MAGICC 450ppm CO₂e stabilisation scenario emissions projections)

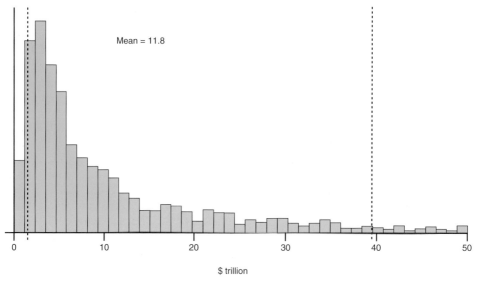

Source: Hope (2008)

In order to cross-check the $12 trillion result, further modelling was undertaken using the very different forest emission projections from the IPCC's SRES A2 scenario.[49] Notwithstanding the large difference in projected emissions, the damage costs were almost identical. The mean NPV of the impacts of BAU deforestation were again around $12 trillion ($1 trillion in 2100; $12 trillion in 2200) – see Figure 2.7b.

As an additional cross-check, the Houghton estimates of deforestation were combined with a global emissions path that reflects a strenuous attempt to limit total CO₂e concentrations. A global emissions path designed to stabilise GHG concentrations at around 450ppm CO₂e was used from the MAGICC model.[50] Again the results were practically identical, with the mean net present value of the impacts of BAU deforestation once more around $12 trillion ($1 trillion in 2100; $12 trillion in 2200) – see Figure 2.7c.

The A2 and 450ppm scenarios cover almost the full range of plausible greenhouse gas emissions pathways over the 21st century. We can therefore conclude that the impacts of BAU deforestation are relatively insensitive to the emissions scenario for other sectors upon which they are superimposed. This is due to a greater forest sector-induced temperature increase from a lower emissions base (such as in the 450ppm scenario), generating similar climate change damage as a lower temperature increase from a higher base (such as in the SRES A2 scenario).

49 Betts et al (2008)
50 Wigley (2003)

2.5 Conclusion

Global forests, particularly tropical rainforests, play a key role in climate regulation. They also provide a range of other benefits such as rainfall for agriculture, flood prevention and biodiversity. However, human activities have significant impacts on forests which in turn affect climate change and wider ecosystem services. Deforestation and forest degradation have particularly detrimental effects on the carbon cycle by releasing large quantities of stored carbon. At the same time, natural, undisturbed forests and ARR have important roles in storing and sequestering atmospheric CO_2.

Rates of global deforestation, particularly in the tropics, are substantial and projected to continue over the next half century. In addition, forest degradation may affect an area the same size or even larger. While global forestation has increased in recent decades, it is not on the scale of deforestation and takes place largely in mid-latitude regions such as China and Europe, where the potential for carbon sequestration is lower than in the tropics.

If the international community is to reach a stabilisation target for atmospheric greenhouse gas emissions that averts the worst impacts of climate change, strong and urgent action will be required to reduce deforestation rates, particularly in the tropics. At the same time, efforts to increase ARR, again particularly in the tropics, will also be important, as will ensuring that standing forests are preserved.

The next chapter examines the drivers of deforestation in different regions of the world to provide an understanding of the economics and policies that will need to change if rates of deforestation are to be reduced. Chapter 4 then goes on to describe a range of more efficient and sustainable management practices that could reduce deforestation while providing better livelihoods for forest communities.

3. The drivers of deforestation

Key messages

World population will probably increase by 50 per cent to 9 billion over the next 40 years. Much of this rise will occur in developing countries. People are getting wealthier – global income per capita could grow by more than 3 per cent per year to 2050, and the global middle class is projected to triple to 1.2 billion.

Demand for agricultural products and timber will continue to rise, increasing already heavy pressure on forest land if productivity increases cannot keep pace. There may be additional pressure directly from population growth in and near forests. Global policy incentives such as biofuels targets could also create pressure for forest clearance, unless effective sustainability criteria are applied.

As long as forest carbon or other ecosystems services are not reflected in the price of commodities produced from converted forest land, forests will – in financial terms – generally be worth more to landowners cut than standing.

The social, institutional and political conditions prevailing in many rainforest nations may amplify the economic pressures on forests. In particular:

- The policy and legal framework in many forest nations is skewed towards deforesting practices, for example through subsidies and tax breaks.
- Lack of clear and secure land tenure is a major factor driving deforestation in many nations. Only when property rights are secure, on paper and in practice, will longer-term investments in sustainable management become worthwhile.
- Weak law enforcement in many countries allows illegal logging to take place on a large scale. It is estimated that, in five of the ten countries with the largest forest cover in the world, more than half of trees cut are felled illegally. Even where policies and laws to help protect forests exist and are clear, many forest nations lack the capacity to implement and enforce them.

Forest transition theory suggests that unless major policy interventions are made on a sustained basis and are effective, then regions with high forest cover will eventually lose a large proportion of forest as a result of economic development.

3.1 Why are trees being cut down?

There already exists much good analysis of the underlying drivers of deforestation.[1] Here we focus on the underlying economic forces. Much of the analysis of deforestation focuses upon developing countries. However, it should not be forgotten that industrialised countries, such as the UK, have already lost most of their tree cover through deforestation over centuries.

Deforestation generally occurs in order to supply timber and agricultural products to meet global and local demand. It will continue until there is sufficient incentive to conserve forests. This will be achieved more easily if the cost of lost forest ecosystems is reflected in the price of the products supplied from the converted land, as discussed in the sections below.

Other drivers play an important role in deforestation:

- policy incentives (whether direct, such as subsidies, or indirect, such as for building roads into forests, or encouraging migration to them);
- land tenure systems;
- other governance-related incentives.

These drivers of deforestation are complex and interlinked. They act globally, regionally and locally. Figure 3.1 illustrates the various types of underlying driver, which are examined in more detail in the rest of this chapter.

Figure 3.1: The underlying drivers of deforestation

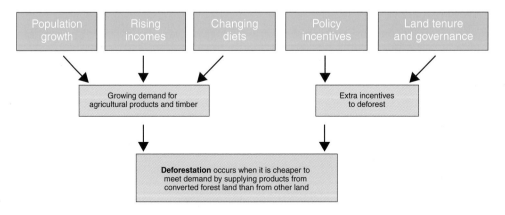

There is considerable variation in the interaction of the various drivers across the different rainforest regions. The Amazon forest is being cleared mainly across a large belt extending from eastern to southern Amazonia. Drivers of deforestation there include expansion of cattle and soybean production. Pockets of cleared forest also occur around settlements and roads. In Africa, deforestation in the Congo Basin is limited, largely because of civil conflict restricting investment and infrastructure expansion. Shifting agriculture tends to be restricted to the secondary forest mosaics and only partially affects the primary

---

1   Such as Kanninen et al (2007)

forest. Illegal logging, urban expansion and fuel requirements are also drivers of deforestation. Most tropical forests in Asia are under intensive exploitation for timber and conversion to agricultural lands, in particular oil palm plantations for the production of vegetable oils. Shifting cultivation is also thought to be on the increase.[2]

3.2 Population growth and wealth creation

World population will probably increase by 50 per cent to 9 billion over the next 40 years. Much of this rise will occur in developing countries. People are getting wealthier – global income per capita could grow by more than 3 per cent per year to 2050, and the global middle class is projected to triple to 1.2 billion.

Figure 3.2 shows projections for global population growth to 2050. Increasing population will place growing pressure on forests by increasing global demand for agricultural commodities and timber products which can be harvested on land that is currently forested. Higher population levels in forest areas themselves may increase pressure directly, through infrastructure expansion.

Figure 3.2: Projections for world population in 2050

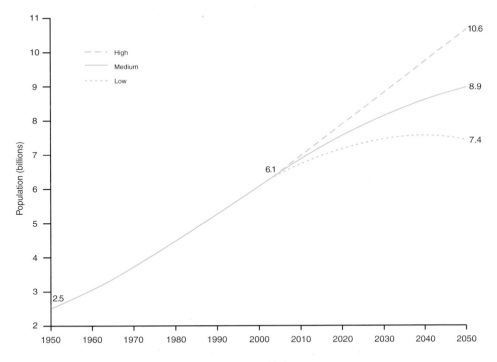

Source: UN Economic and Social Affairs Department (2004)

[2] Betts et al (2008)

Even faster than the growth in the world population is the growth of the global middle class, as a result of economic development. The World Bank estimates that by 2030, 1.2 billion people in developing countries – 15 per cent of world population – will belong to the global middle class,[3] up from 400 million in 2005 (see Figure 3.3).

Figure 3.3: Growth in global middle class

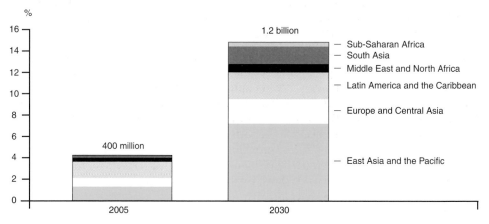

Source: World Bank (2007)

This growing global middle class is important for understanding changing demand because it is able to afford more meat and dairy produce. This kind of diet requires much more land to support it compared with a vegetable-based diet (Table 3.1).

Table 3.1: Land requirements for producing different types of food

Food item	Land requirement (m² year kg⁻¹)
Beef	20.9
Cheese	10.2
Pork	8.9
Eggs	3.5
Flour	1.6
Whole milk	1.2
Fruits (average)	0.5
Vegetables (average)	0.3
Potatoes	0.2

Source: Gerbens-Leenes et al (2002)

---

3   A family of four in that class earns between $16,000 and $68,000 in purchasing power parity (PPP) dollars.

3.3 Growing demand for agricultural products and timber

Demand for agricultural products and timber will continue to rise, increasing already heavy pressure on forest land if productivity increases cannot keep pace. There may be additional pressure directly from population growth in and near forests. Global policy incentives such as biofuels targets could also create pressure for forest clearance, unless effective sustainability criteria are applied.

Global demand for agricultural products has been increasing for many decades. Yet, in real terms, prices of most agricultural commodities have fallen over the last 30 years, primarily as a result of technology-driven improvements in productivity and yields. Most increased agricultural production has come from increased yields, rather than greater use of land.[4] However, despite the declining returns for agricultural products, deforestation has continued, at least in part because the costs of ecosystem loss have not been reflected in the price of products grown on converted forest land. On the contrary, sales of timber from deforesting the land often partly or totally subsidise the conversion to agriculture. And the construction of more roads (both legal and illegal) has brought down the cost of sending produce to market considerably.

Since 2006 there has been a strong rise in food prices, which has occurred simultaneously for all key food staples. Contributing factors include:[5]

- a rapidly growing global middle class (see above);
- increasing demand for biofuels;[6]
- poor harvests as a result of more extreme weather conditions;
- export restrictions imposed in response to higher commodity prices.

In the medium term, global food prices are expected to resume their decline in real terms, albeit at a slower rate, due to increased supply responding to higher prices and improved weather conditions.[7]

A direct relationship can be shown between the sale price of agricultural commodities and deforestation (see, for example, Figure 3.4 below).

---

4 HM Treasury (2008)
5 HM Treasury (2008)
6 Gallagher (2008)
7 OECD-FAO (2008)

Figure 3.4: Correlation between deforestation in the Brazilian Amazonia, farmgate prices of beef and rainfall (2001-2003)

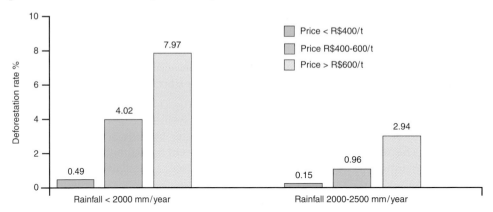

Source: Chomitz et al (2006)

The factors set out above will mean that demand for agricultural products and timber will continue to increase over the decades to come (see Figure 3.5 for increases in demand over past years). The past half century has seen dramatic improvements in agricultural yield, but future productivity increases may not be enough to keep up with demand, particularly if biofuel is in competition with food production for land.[8]

Figure 3.5: Demand for major food crops (difference from previous year in millions of tonnes)

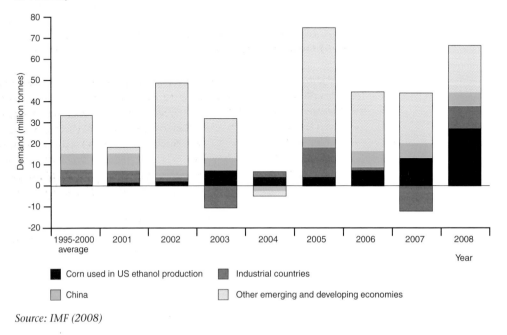

Source: IMF (2008)

8 Gallagher (2008)

3.4 Current economic incentives for landowners to deforest

As long as forest carbon or other ecosystem services are not reflected in the price of commodities produced from converted forest land, forests will – in financial terms – generally be worth more to landowners cut than standing.

Table 3.2 below gives some examples of land use returns that can be derived from converted forest land. As previously mentioned, timber sales are often used to fund the costs of conversion.

Table 3.2: Selected land use returns in some forest nations

Country	Land use	Land use returns ($/ha)
Brazil	Soybeans	3,275
	Beef cattle (medium/large scale)	413
	One-off timber harvesting	251
	Beef cattle (small scale)	3
Indonesia	Large scale palm oil	3,340
	One-off timber harvesting	1,099
	Smallholder rubber	72
	Rice fallow	28
Cameroon	Cocoa with marketed fruit	1,448
	Annual food crop, short fallow	821
	Annual food crop, long fallow	367

Note: Returns are net present value in 2007 $ at discount rate of 10 per cent over 30 years
Source: Grieg-Gran (2008)

Even if forests are sustainably harvested, the revenues available are often unattractive compared with those available from conversion. In one Brazilian example,[9] reduced impact logging would yield $128 profit per hectare from an initial selective harvest. If the forest were left alone to regenerate, another harvest would be possible in 30 years. This gives a net present value of only $0.24 per hectare,[10] which is very unattractive compared with conversion to one of the land uses in Table 3.2.

---

9   Boltz et al (2001), discussed in Chomitz et al (2006)
10 A 20 per cent discount rate was used in this instance.

3.5 Policy incentives

The social, institutional and political conditions prevailing operating in many rainforest nations may amplify the economic pressures on forests. The policy and legal framework in many forest nations is skewed towards deforesting practices, for example through subsidies and tax breaks.

Policy incentives, both direct and indirect, are important drivers of deforestation (Figure 3.6). Policies and laws governing land use set the context for whether sustainable management of land and forests will be feasible, effective, profitable and desirable – or if demand will be met through deforestation. Land use policies and forest law systems in many forest nations have so far proved ineffective at preventing forest destruction, and in many cases, have favoured deforestation. The analysis in the previous section demonstrates why such policies have often made good economic sense for many countries.

Figure 3.6: Policy drivers of deforestation

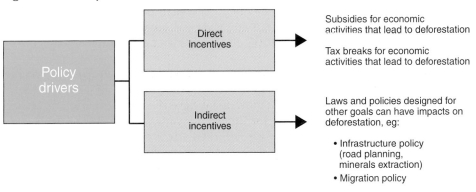

Although sustainably managed today, economic development and industrialisation over several hundred years greatly depleted Europe's forests. As populations grew and farming methods changed, significant areas of forest were lost to livestock grazing, swidden cultivation and clearance for permanent fields. Remaining areas were subject to intensive logging to meet the needs of local populations and for industries such as shipbuilding, construction, charcoal production and mining.[11] Only about 0.24 per cent of Europe's remaining 144 million hectares of forest is considered to be virgin forest and only 1.8 per cent is classified as virgin forest or old growth forest remnants.[12]

The US has also experienced significant deforestation, having lost 90 per cent of the virgin forests that once covered the lower 48 states.[13] So, until very recently, deforestation made economic sense for most developed countries, and state policies and processes supported this. In seeking to help reduce deforestation in tropical forest nations, developed countries can therefore be seen as asking developing countries to reject a high-deforestation (and consequently high-carbon) growth path that they themselves have followed.

---

11 Colchester (1998)
12 Colchester (1998)
13 University of Michigan (2006)

Many forest nations have also pursued development and growth policies, in both the forestry and non-forestry sectors, which have led to higher levels of deforestation. Examples of this include tax breaks and subsidies for economic activities that result in deforestation. Sentiano reports on how policies pursued in Indonesia following the country's mid-1990s economic crisis led to massive overcapacity in the timber processing sector, with annual demand for wood-based industries far outstripping the legal and sustainable timber supply.[14] Many South American countries have supported the development of export sectors such as sugar and beef in an attempt to generate foreign exchange earnings,[15] leading to forest clearance to make way for pasture and crops.

Furthermore, legal and policy frameworks have tended to favour industrial interests in timber, agriculture and mineral extraction, over the interests of forest-dependent communities in managing land and resources,[16] and this can make overexploitation of forest resources more likely. Evidence suggests that ownership and control over forests by indigenous communities, within an appropriate framework of regulation and support, can limit deforestation.[17] Local communities – through their local knowledge and expertise, their role in protecting forests from outside encroachment and a long-term commitment to their lands – often exercise superior conservation practices.[18] However, policies skewed towards larger industrial interests disadvantage smaller and community-based enterprises and reduce their profitability;[19] and large timber companies tend to receive a greater share of concessions.

The indirect effects of non-forestry policies can also be significant, if environmental impacts are not taken into account. Infrastructure development of all kinds can lead to forest clearance. Road building and improvement have the biggest impact on deforestation. They make transport cheaper for timber companies, encourage expansion and infrastructure development at the forest frontier and drive cycles of conversion of timber harvesting and conversion to agriculture.[20] Migration to settlements, and the establishment of new ones in forest frontier areas, can also reinforce economic incentives for further deforestation through the growth of local markets and infrastructure, which further attracts new migrants.[21]

---

14 Sentiono (2007)
15 Lambin and Geist (2003)
16 Sunderlin et al (2008)
17 Chomitz et al (2006)
18 Molnar et al (2004)
19 Molnar et al (2006)
20 Kanninen et al (2007); Chomitz et al (2006)
21 Lambin and Geist (2003)

3.6 Land tenure

Lack of clear and secure land tenure is a major factor driving deforestation in many nations. Only when property rights are secure, on paper and in practice, will longer-term investments in sustainable management become worthwhile.

Lack of clear and secure land tenure and use rights is a major factor driving deforestation in many forest nations. Table 3.3 shows the distribution of nominal forest tenure in four key countries. There is a wide variation in land tenure models between forest nations, from complete state administration of forest lands in much of central Africa, to almost 100 per cent private ownership by communities and indigenous groups, as in Papua New Guinea. Global trends show an overall decline in the area of forest under state ownership and corresponding increases in the area of forests designated for use by, or owned by, communities and indigenous peoples, or by private individuals and firms. [22]

Table 3.3: Forest tenure and distribution (million ha)

Country	Public		Private	
	Administered by government	Designated for communities/ indigenous peoples	Communities/ indigenous peoples	Individuals and firms
Brazil	88.6	25.6	109.1	198.0
Indonesia	121.9	0.2	0	1.7
DR Congo	133.6	0	0	0
Papua New Guinea	0.3	0	25.5	0

Source: Sunderlin et al (2008)

Given the potential advantages of community ownership and stewardship of forest lands, this could be seen as a positive trend from the viewpoint of managing forests sustainably to tackle climate change. In reality, a lack of clarity and security of land tenure and user rights is extremely widespread in forest nations, and this acts as a key underlying driver of deforestation. Only when property rights are secure, on paper and in practice, do longer-term investments in sustainable management become worthwhile.[23] Governments, communities and individuals can successfully manage forests in a sustainable way, but only when the appropriate enabling elements are in place. Secure tenure rights and resource access are two of these elements.[24]

The lack of clear and secure tenure is a feature of the broader political and economic marginalisation of forest communities, who often struggle to assert and exer-

---

22   Sunderlin et al (2008)
23   Kanninen et al (2007)
24   Molnar et al (2004)

cise their rights over forest land and use forest resources as they wish. In the confusion and conflict over land rights, it is often indigenous people and forest communities who are disadvantaged. Concessions are often awarded on lands that have been designated for use by, or titled to, indigenous peoples. In Peru, for example, almost all titled indigenous lands are affected in some way by the 45 million hectares of land under contract for oil and gas exploration and exploitation.[25] Rights and Resources Initiative reports that, in Colombia, paramilitary groups have forcibly evicted forest peoples and sold their lands to speculators and for palm oil plantations. The Indonesian non-governmental organisation Sawit Watch reports that at least 400 communities in Indonesia have been affected by land conflicts caused by the expansion of palm oil plantations.[26]

3.7 Capacity

Weak law enforcement in many countries allows illegal logging to take place on a large scale. It is estimated that, in five of the ten countries with the largest forest cover in the world, more than half of trees cut are felled illegally. Even where policies and laws to help protect forests exist and are clear, many forest nations lack the capacity to implement and enforce them.

Poor regulations and a lack of implementation and enforcement capacity in many forest nations are widely recognised as important underlying drivers of deforestation. As Figure 3.7 shows, high levels of forest loss, or net deforestation, tend to be correlated with lower government effectiveness, based on World Bank governance indicators. Governance weaknesses related to deforestation also overlap with the underlying drivers of poverty in developing countries and are particularly relevant to forest communities.

Ambiguous or overlapping laws, regulations and jurisdictions provide opportunities to exploit 'grey areas' and circumvent forest protection policies.[27] Complicated devolution processes, for example, can result in contradictory and inconsistent legal frameworks, open to exploitation by actors seeking quick profits through deforesting activities.

Even where policies and laws are clear, forest nations may lack the capacity to implement and enforce them. Weak law enforcement in rainforest nations allows illegal logging to take place on a large scale globally. It is estimated that, in five of the ten countries with the most forest, more than half of the trees cut are felled illegally.[28] Weak institutional capacity at all levels significantly hampers the ability of countries to implement forest protection programmes and promote sustainable management. Ministries with responsibility for different aspects of land use may not be well coordinated. This may reflect in part budgetary constraints, resulting from competing priorities for funding in developing countries.

Forest resources, like other natural resources, are subject to capture and exploitation by political, economic and military elites. A lack of transparency in decision-making in some forest nations further enables powerful political and corporate interests to act with

---

25 Sunderlin et al (2008)
26 Sunderlin et al (2008)
27 Kanninen et al (2007)
28 Brack (2007)

minimal public accountability,[29] which will tend to lead to unsustainable practices.

Figure 3.7: Correlation between indices of government effectiveness and net deforestation in 40 forest nations

Note: Each point represents a forest nation. The world's 40 largest forest countries, based on area of forest

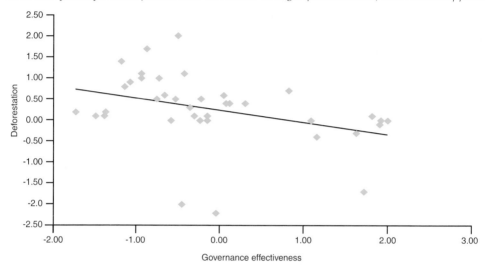

cover, were included in the analysis. Governance index scores range from -2.5 (less effective) to +2.5 (more effective).
Source: FAO (2005). Government Effectiveness index scores from World Governance Indicators for 2000, World Bank (2008)

The relative importance of these governance factors – policy, capacity and tenure issues – as drivers of deforestation varies from country to country. Lambin and Geist identified the main policy and capacity factors underlying deforestation in the three main rainforest regions. These are set out Table 3.4.

Table 3.4: Governance drivers of deforestation in the three major rainforest regions

	Latin America	Southeast Asia	West and Central Africa
Governance drivers	Policies facilitating land transfer to large, private ranches; state policies of frontier colonisation	Policies facilitating colonisation and state plantations; large transmigration projects; weak enforcement of forestry law; insecure land ownership	Poor law enforcement; mismanagement by weak nation states; in-migration

Source: Lambin and Geist (2003)

---

29 Kanninen et al (2007)

3. The drivers of deforestation

3.8 Forest transitions over time

Forest transition theory suggests that unless major policy interventions are made on a sustained basis and are effective, then regions with high forest cover will lose a large proportion of forest as a result of economic development.

The combined impact of economic and other drivers on forests over time is described by the forest transition concept. This sets out a sequence of events in which forest cover first declines then reaches a minimum level before slowly increasing and eventually stabilising. This sequence is illustrated in Figure 3.8 below:

Figure 3.8 Stages and main drivers in forest transitions

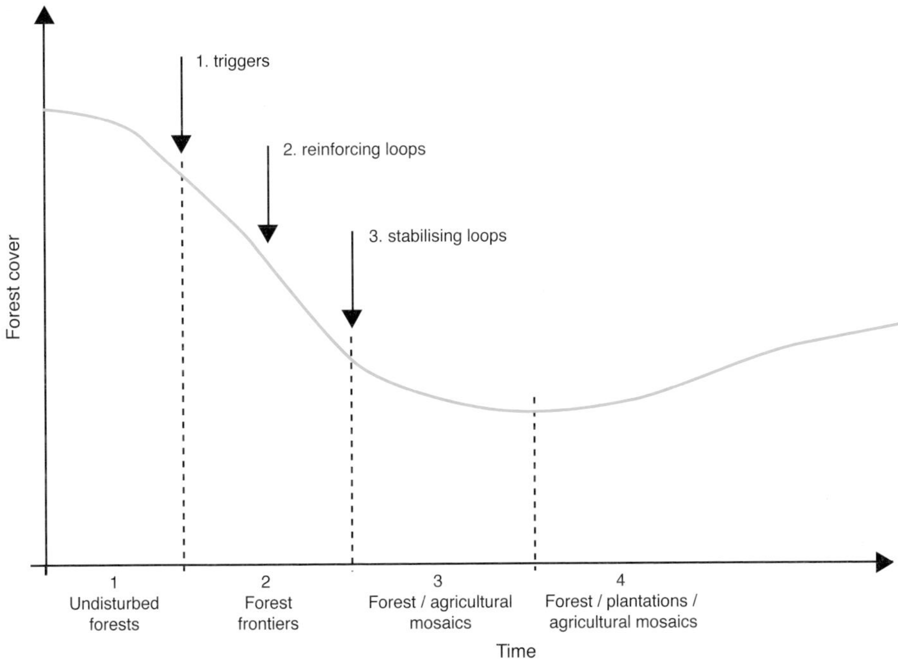

[Graph showing forest cover over time with four stages: 1. Undisturbed forests, 2. Forest frontiers, 3. Forest / agricultural mosaics, 4. Forest / plantations / agricultural mosaics. Arrows indicate: 1. triggers, 2. reinforcing loops, 3. stabilising loops]

Source: Angelsen (2007)

Movement along the forest transition curve is the result of three sets of forces:

- A set of *triggers* (force 1) initiates the deforestation process. The key trigger is the construction of new or improved roads. Formerly, relatively undisturbed forest would have enjoyed passive protection due to poor infrastructure and market access. But now the area is opened up to both people and capital, and a market outlet is created for agricultural and timber products.
- Next a set of *reinforcing loops* (force 2) enlarge the initial effect through positive feedback. Population and economic growth in the area increase agricultural rent through increased local demand and the development of downstream processing activities for agricultural products. Agricultural inputs become cheaper as capital accumulates and the population in the area expands. And population and economic growth in the area

also stimulate the development of better infrastructure and transport facilities, further reducing transport costs and increasing agricultural rents.
- Eventually, *stabilising loops* result in downward pressure on agricultural rent and an upward shift in forest rent. The downward pressure on agricultural rent occurs as better-off farm wages and employment opportunities attract people out of the agricultural labour market. Increased agricultural supply may also dampen prices. Economic development can also lead to a demographic transition that reduces the supply of labour and drives up wages. Forest rents also tend to rise in parallel: high levels of deforestation lead to forest scarcity, which leads to higher prices for forest products. This encourages both better forest management and reforestation. Loss of forest ecosystem services may also result in policy changes that promote forest conservation. These forces are amplified by economic development that brings about an improved rule of law and greater political awareness of environmental services.

Forest transition theory suggests that unless major policy interventions are made on a sustained basis and are effective, then regions with high forest cover will eventually lose a large proportion of forest as a result of economic development. Although cover will eventually be partially restored, it will not have the same carbon storage capacity or biodiversity as the old growth forest that was there before. Furthermore, by the time the global forest sector sequesters more CO_2 than it emits, a climate tipping point might have already been reached.[30]

3.9 Conclusion

Many developed countries are already largely deforested. Deforestation is now principally happening in developing countries as they tread a similar path to meet their development needs.

With an increasing and increasingly wealthy global population, demand for land and for agricultural and forest products will continue to rise. Without a value on the local and global services provided by standing forests, there is little economic incentive to meet this demand other than through deforestation. And current policy incentives in many countries amplify the global and local economic incentives to deforest. Weak governance, particularly in forest nations, further contributes to continued unsustainable levels of deforestation. Later chapters in this Review will look at how bringing forests into a global deal on climate change can help shift the global response to demand away from deforestation and towards more efficient and sustainable land and resource use.

---

30 Angelsen (2007)

4. Sustainable production and poverty reduction

Key messages

Population growth and wealth creation are increasing demand for agricultural products and timber. Clearance of forested land is currently meeting the growing demand for these commodities.

A significant land gap for meeting demand for commodities is a real threat, which will put increased pressure on forests. More research is urgently needed on land availability at national and sub-national levels. However, it is clear that if deforestation is to be reduced, a global step-change is needed in the way land is used and commodities are produced.

Our vision is a sustainable system of global production which can meet increasing demand for commodities and lead to reduced carbon emissions, better livelihoods for the poor and preservation of non-carbon ecosystem services such as biodiversity and water services.

To achieve this vision, a shift of policies and practices in several sectors will be required:

- Production of agricultural commodities can be made more efficient and sustainable mainly through increased productivity. Well-managed expansion onto non-forest land and agroforestry will also have a role.
- Global adoption of sustainable forest management is required to meet timber needs. The role of communities as environmental stewards will be particularly important.
- Infrastructure expansion needs to be managed in a way that minimises environmental impacts and benefits local populations. Partnerships between companies and communities could help make this happen.
- The promotion of off-farm employment opportunities, as part of a broader economic strategy, could help reduce deforestation.
- Protected areas are likely to have a significant role in preserving global forests, but their design and management require the full participation of affected communities.
- Payments for ecosystem services could provide a complementary income stream to that generated by sustainable production.
- Certified products and effective biofuels sustainability criteria that include the indirect effects of land use change can also support the shift to sustainable practices.

Three key levers can help make the shift from deforestation to more sustainable policies and practices: valuing carbon and other services that forests provide; shifting policy incentives to more sustainable and efficient production practices; and using demand-side measures to support sustainable production.

4.1 Introduction

Population growth and wealth creation are increasing demand for agricultural products and timber. Clearance of forested land is currently helping meet the growing demand for these commodities and a significant land gap for meeting this demand is a real threat. Although a detailed discussion of land availability for agricultural use is outside the scope of this Review, we briefly outline the challenge in this chapter.

While the scale of the land gap is uncertain, a step-change is clearly needed in the way land is used and commodities are produced. This chapter sets out the Eliasch Review's vision for a sustainable and efficient response to global demand for commodities. It then identifies three key levers that can be used to make the shift from deforestation to sustainable production: valuing carbon and other services that forests provide, shifting policy incentives to more sustainable and efficient production practices, and using demand-side measures to support sustainable production.

4.2 Land availability

Population growth and wealth creation are increasing demand for agricultural products and timber. Clearance of forested land is currently meeting the growing demand for these commodities.

A significant land gap for meeting demand for commodities is a real threat, which will put increased pressure on forests. More research is urgently needed on land availability at national and sub-national levels. However, it is clear that, if deforestation is to be reduced, a global step-change is needed in the way land is used and commodities are produced.

Concerns over the impacts, including on forests, of the rapid expansion of biofuels to meet US and EU fuel targets led to the UK government commissioning the independent Gallagher Review in 2008.[1] The review developed projections for land availability and land requirements to meet demand for commodities to 2020. The findings highlighted the extent of this uncertainty and the need for further analysis. Figure 4.1 shows the Gallagher Review's projections, under different scenarios of global land availability to meet growing consumer demand for commodities and biofuel targets.

---

1 Gallagher (2008)

Figure 4.1: Projections of global land availability to meet growing consumer demand for commodities

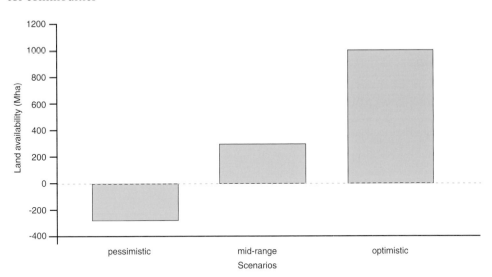

Source: Gallagher (2008)

The optimistic scenario, which assumes some of the significant developments in technology and production methods for agricultural production discussed later in this chapter, suggests that there is no cause for concern and that there will still be more than 900 million hectares of productive land available in 2020. The mid-range scenario suggests a surplus of land of more that 200 million hectares. The pessimistic scenario sees a land gap of approximately the same size. Gallagher also draws attention to the variety of estimates of demand for timber, which are excluded from the above scenarios, including the FAO's projections that land under forest plantation could stay static at the current 190 million hectares or rise to as much as 310 million hectares by 2020. Alternative global land availability analysis by Roberts and Nilsson estimated a likely land gap of at least 215 million hectares by 2030.[2]

All land availability estimates contain considerable uncertainties and are based on different, complex and sometimes speculative assumptions. The amount and type of non-forested land available also varies significantly between countries and regions, and drivers other than national-level land pressure may be the main factor in the encroachment of farmland onto forest land in many areas. More work is needed to understand the extent of the problem.

Even without a land gap, the central and optimistic cases presented by Gallagher assume considerable productivity improvements in agriculture beyond current baselines. Consequently, in all cases, meeting the challenge of matching land availability with demand for land and commodities while reducing deforestation requires a step-change in the way land is used and commodities are produced.

---

2   Roberts and Nilsson (2008)

4.3 A vision of sustainable production

Our vision is a sustainable system of global production which can meet increasing demand for commodities and lead to reduced carbon emissions, better livelihoods for the poor and preservation of non-carbon ecosystem services such as biodiversity and water services.

Given the uncertainties about land availability over the coming decades, any system to reduce deforestation will need to address rising demand for commodities. This Review's long-term vision is a system of sustainable and efficient production to meet demand for commodities that leads to reductions in carbon emissions, better livelihoods for the poor and preservation of other ecosystem services such as biodiversity and water services. Figure 4.2 outlines the elements of this vision.

Figure 4.2: A vision of sustainable production

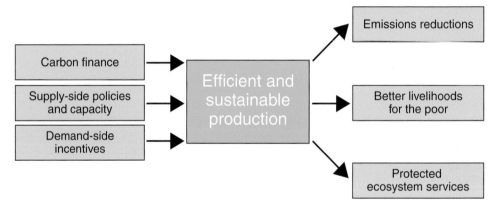

The response to increasing demand for commodities and land currently leads to deforestation. Our vision sees three main levers coming into play to make the shift to sustainable production. These levers, on the left of the diagram, are:

- valuing carbon and other services that forests provide;
- shifting policy incentives to more sustainable and efficient production practices;
- using demand-side measures to support sustainable production.

Section 4.7 describes the respective roles of these levers.

The effectiveness of mechanisms and programmes intended to change the way demand for land and commodities is met will depend on how well they contribute to the development needs of forest-dependent communities. It is estimated that 1.6 billion people depend to some degree on forests for their livelihoods.[3] Changing patterns of land use will impact on the interests and livelihoods of hundreds of millions of people, many of them among the world's poorest. While there may be trade-offs between environmental and social goals in the short term, long-run sustainability means that

---

3   World Bank (2004)

new models of land use will need to benefit poor people and forest communities.

Forests and the ecosystem services they provide can also play a significant role in adaptation to climate change. More resilient ecosystems will be better able to withstand shocks and support communities in adapting to the effects of climate change.[4] Programmes to reduce deforestation that take a holistic approach to mitigation, adaptation and livelihoods could therefore be a particularly efficient and equitable response to climate change.

Forests provide a huge range of other ecosystem services, in addition to being a carbon store and sink. They regulate water supply used for agriculture nearby. They also support considerable biodiversity and play an important role in global atmospheric circulation. Forest use that damages the forest structure not only produces carbon emissions but also damages the ecosystem services they provide. Therefore, methods that reduce CO_2 emissions can preserve other forest ecosystem services that are crucial to maintaining life and livelihoods.

4.4 Sustainable production and conservation

To meet the vision of sustainable production, a shift of policies and practices in several sectors will be required: agriculture, timber and wood products, infrastructure, alternative employment, and forest conservation.

Forest nations will want to use country-specific policies to reduce deforestation, but in all cases a coherent cross-government policy, based on national and sub-national analysis of the drivers of deforestation, will be essential. Chapter 12 examines the role of national governments in achieving the shift to sustainable production. What follows is an indicative sample of some of the on-the-ground approaches forest nations may choose to adopt to shift production onto a more sustainable footing.

4.4.1 Agricultural production

Production of agricultural commodities can be made more efficient and sustainable mainly through increased productivity. Well-managed expansion onto non-forest land and agroforestry will also have a role.

Agricultural intensification has significant potential to reduce pressure on forest land by meeting demand more efficiently. Productivity for cereals (in tonnes per hectare) has been stagnant in sub-Saharan Africa for around 30 years, whereas cereal yields in East Asia rose by 2.8 per cent a year in 1961–2004.[5] Analysis for Gallagher's review[6] found that 50-70 million hectares of pasture land could be made available in Brazil if the productivity rate in São Paulo was extended to the rest of the country.[7] There is a range of technologies that could improve incomes and environmental outcomes, such as technological and bioengineering improvements (including new generation biofuels) and improved pest

---

4 Corbera (2007)
5 World Bank (2008)
6 Gallagher (2008)
7 Volpi (2008)

control measures.[8] Agricultural intensification has the potential to help increase global food security and support livelihoods, as well as reduce pressure on forests.

The challenges are considerable, however, and significant development and wider diffusion of agricultural technologies appropriate for rainforest nations will be required. Achieving intensification while maintaining biodiversity is a further challenge, although good practice examples exist.[9] And while the biggest impact of this type of approach could be around central African rainforests where forest degradation from smallholder cultivation is the major threat to forests, this is also where there is least capacity to implement large-scale programmes. Linking agricultural programmes to forest protection, and possibly carbon finance, would also require far more integrated consideration and planning of land use at national and local levels.

In addition, although the effect of intensification would be to reduce the average pressure on global forests, if intensification increased agricultural profitability, then the pressure to deforest could increase in the areas where the more intense practices applied. The prospect of intensification therefore increases the importance of proper valuation of carbon and other ecosystem services.

Agricultural extensification onto non-forested land not currently being used for agriculture, may also offer possibilities in some areas. The Brazilian Cerrado region, for example, has an estimated 106 million hectares of currently unused land which would be suitable for agriculture, outside forested land.[10] Estimates also indicate that there are at least 16 million hectares of lands which were converted to agriculture and cattle ranching in the Brazilian Amazon and have now been abandoned.[11] The Gallagher Review also looked at the potential for policies that promote biofuels expansion onto land not currently being used for agriculture (though its use by landless people for subsistence purposes may be invisible to public authorities) and marginal or degraded land not suited for food production or degraded through deforestation. The potential to pursue this option will depend on local circumstances and again there is considerable uncertainty over the availability and suitability of these kinds of lands.

Agroforestry systems, in which trees are interspersed across pasture and cultivated land, can be one way to achieve the combined benefits of improving income streams from agriculture, protecting biodiversity and maintaining or increasing forest cover. As with agricultural intensification, the technical and capacity building challenges are significant, though successful examples are available. Box 4.1 describes the Plan Vivo approach, in which carbon finance provides supplementary income to sustainable forest management/agroforestry practices which themselves aim to improve livelihoods.[12] The RISEMP[13] scheme in Colombia, Costa Rica and Nicaragua aims to use payment for ecosystem services to catalyse adoption of silvopastoral systems by paying landowners for beneficial changes in land cover. After two years, the programmes had tripled the share of the project areas considered to be 'improved pasture with high tree density'.[14]

---

8   Chomitz (2006)
9   World Bank (2008)
10 Brazilian Ministry of Agriculture
11 Brazilian Government (2004)
12 Regional Integrated Silvopastoral Ecosystem Management Programme
13 www.planvivo.org
14 Chomitz et al (2006)

> **Box 4.1: Scolel Te: Plan Vivo**
>
> Plan Vivo programmes aim to:
>
> - sequester carbon through forest and agricultural practices which contribute to sustainable livelihood systems;
> - assist farmers and communities to develop more sustainable land management and better livelihoods through the provision of carbon services;
> - target low-income farmers who often live in marginal areas, bringing together smallholders and communities to deliver benefits in the markets for environmental services.
>
> The Scolel Te programme in Southern Mexico includes over 2,000 families of indigenous Mayan and Mestizo farmers in 30 communities. The programme provides support to develop sustainable forestry and agroforestry techniques to improve livelihoods. It includes supplementing landholders' income with carbon finance from offsets sold on voluntary carbon markets. It has the potential to sequester around 100,000tCO_2 per year.
>
> Several forestry systems are used in the Scolel Te project to sequester carbon:
>
> - the establishment of tree plantations on areas previously used as pasture may increase carbon stored in vegetation by about 440tCO_2/ha;
> - by growing timber and fruit trees interspersed with annual crops such as corn or perennial crops such as coffee, around 256tCO_2/ha can be sequestered;
> - where closed forests are threatened, protection can prevent emissions of up to 1100tCO_2/ha; and where forests are degraded, careful management and restoration can increase carbon storage by around 440tCO_2/ha.
>
> More than $30,000 in carbon payments was made to Scolel Te producers across around 20 communities in 2006.

4.4.2 Timber and non-timber forest production

Global adoption of sustainable forest management is required to meet timber needs. The role of communities as environmental stewards will be particularly important.

Sustainable forest management (SFM) aims to maintain and enhance the economic, social and environmental values of all types of forests, for the benefit of present and future generations.[15] The potential benefits of SFM are considerable: simultaneously contributing to adaptation (through maintaining resilient ecosystems), climate change mitigation and poverty reduction.

A study commissioned for this Review concluded that the impact of deforestation on carbon stocks is often an indicator of damage to other ecosystem services, and so forest management to minimise carbon emissions may also maintain many other forest ecosystem benefits. The practices that minimally disturb, replace or maintain the original

---

15 United Nations Economic and Social Council (2007)

structure of primary tropical forests tend to be those that are most likely to be sustainable in the long term. Examples include agroforestry, reduced impact logging (which can lead to a gain of as much as 50 per cent in the 'carbon stocks' from remaining vegetation)[16], conservation and regeneration/rehabilitation. However, the long-term sustainability of these activities will depend heavily on forest planning, monitoring and adaptive management strategies to ensure the successful maintenance of the ecosystem services.[17]

Plantations can also be used to meet demand for wood. Planting extensive monocultures of non-native tree species can increase carbon stocks while providing little support for local biodiversity, so the impact on ecosystem services will again depend on the planning and management techniques used.[18]

Sustainable forest management is a stated national policy objective of many forest nations, but achieved by relatively few. Less than 5 per cent of global tropical forest area is considered to be sustainably managed.[19] Investment in forestry has mostly taken a short-term perspective. Companies and communities exploiting forests need to see the economic returns from SFM for it to become widely viable.

Community forest management is a model for SFM which recognises forest communities' comparative advantage as environmental stewards and their strengthening political voice.[20] There is evidence that community forest management, where successfully applied, has reduced deforestation, generated more sustainable income streams for communities and contributed to the acquisition of technical skills. In some cases it has also led to greater transparency in decision-making.[21]

Communities may face particular entry barriers to playing their part in SFM however. Policies and subsidy schemes to encourage SFM have usually been designed with large formal industry in mind. Regulatory frameworks are often slow and costly to negotiate.[22] Many communities also need considerable and long-term technical, management and administrative support to take advantage of community forest mechanisms.[23]

Box 4.2 describes two community forest management set-ups that the Review Team noted in Brazil and the Congo Basin.

---

16 CIFOR (1998)
17 Sajwaj et al (2008)
18 Sajwaj et al (2008)
19 ITTO (2006)
20 White and Markin et al (2004)
21 Moss et al (2005), Chomitz (2006)
22 Molnar et al (2006)
23 Chomitz et al (2006)

Box 4.2: Community forest management: two examples

Projeto Ambé, National Forest of Tapajós, Santarém, State of Pará, Brazil
The National Forest of Tapajós comprises 600,000 hectares of forests on the right margin of the Tapajós river in the region of Santarém, State of Pará. In 2006, Projeto Ambé, an initiative on sustainable community forestry on 32,000 hectares of the National Forest, began with support from PPG7 (Pilot Program to Conserve Brazilian Rainforest). This is the largest initiative in Brazil on community forestry. COOMFLONA (Cooperativa Mista Flona Tapajós Verde), a cooperative involving 132 families from communities in the National Forest of Tapajós, organises the project. In 2007, timber harvest in 400 ha of forests produced about 5,000m^3 of timber, from 60 different species. COOM-FLONA sold the timber, investing 50 per cent in the 2008 production and 15 per cent in community projects within the National Forest; 20 per cent was shared among cooperative partners; 10 per cent was destined to a reserve fund and 5 per cent to a fund for technical, educational and social assistance. In 2008, 13,497m^3 of timber has been produced by communities. An auction of timber took place in September 2008. At the time of going to press, communities expected to sell the whole set of timber for R$3.1 million (around $2 million).

Congo Basin – sustainable timber harvesting
The Review Team saw an example of sustainable timber production in the Congo Basin where an international non-governmental organisation had provided up-front costs and produced the plan required to obtain a community logging licence. The community was allowed to harvest 1400m^3 of timber per year, but had only achieved 40m^3 in its best year. The economic contribution was small and ownership of the scheme was limited. This failure to reach its potential was a result of bureaucratic obstacles (it had to reapply every year for a logging licence, and this was delayed for several months), and a lack of equipment and expertise.

Several interlocutors the Review Team met warned against a too simplistic understanding of 'community'. There is rarely homogeneity between or within communities, and they may neither be organised nor have an agreed representative to engage with external processes and organisations. Thus, although the potential for community forest management in reinforcing communities' rights, improving livelihoods and contributing to climate change mitigation is clear, it is by no means a quick fix.

4.5 Infrastructure and alternative employment

4.5.1 Infrastructure expansion and other industries

Infrastructure expansion needs to be managed in a way that minimises environmental impacts and benefits local populations. Partnerships between companies and communities could help make this happen.

Avoiding negative environmental impacts from infrastructure expansion requires that climate change, deforestation and livelihoods considerations are mainstreamed into national growth and development strategies. The rigorous application of environmental and social impact assessments to all major policy developments, particularly road building, will be a key means for governments to expose the inevitable trade-offs between different policy objectives, make decisions in the full knowledge of the likely impact on deforestation and rural livelihoods, and put in place mitigation strategies where necessary.

Companies establishing new infrastructure can work with government and communities to ensure they make a net positive contribution to reducing emissions, promoting local livelihoods and enhancing other ecosystem services.

For example, Conservation International's guide to responsible large-scale mining, Lightening the Lode, suggests how mining companies could provide 'financial or in-kind support for management of the national park system, support for research scientists, participation in the creation and management of a new local protected area or indigenous reserve, or contributions to local governmental or non-governmental conservation and community development programs'.[24]

These types of partnership are relevant for all activities where industrial production could threaten forests and communities if safeguards are not put in place. In the Northern Republic of Congo, for example, the Wildlife Conservation Society is working with logging company Congolaise Industrielle des Bois (CIB) to put in place more sustainable practices and ensure that local biodiversity is protected in areas around the logging concession.[25] The Nature Conservancy (TNC) is working with 210 soy farmers to help them develop environmentally sustainable practices that conform to Brazilian environmental legislation. TNC estimates that this programme has the potential to conserve nearly 1.2 million acres of tropical forest.[26] Full evaluation of the impacts of these types of programmes, including of the extent to which they can provide benefits for companies and communities, will be critical to feed into the development of future partnership programmes.

4.5.2 Promotion of off-farm employment

The promotion of off-farm employment opportunities, as part of a broader economic strategy, could help reduce deforestation.

As demand for agriculture and timber products continues to grow, the need for labour to produce them will continue. In some areas, however, deforestation from subsistence

---

24 Sweeting and Clark (2000)
25 www.itto.or.jp
26 www.tnc.org

farming may occur through a lack of alternative livelihoods for those living in and near forests. In such areas the promotion of industries generating off-farm employment opportunities may help to reduce deforestation. What is possible and appropriate will depend on the development trajectory of countries and regions, and will form part of a broader strategy. Increases in agricultural productivity may be a pre-condition for the generation of off-farm opportunities, for example. Box 4.3 describes one vision of the development of technology centres in the Amazon, which could drive growth and reduce pressure on Brazil's rainforests.

> **Box 4.3: A scientific and technological revolution for the Brazilian Amazon**
>
> A number of proposals have been put forward for shifting regional economies from land use to service sectors. Below are extracts of an article setting out one vision for the development of technology centres in the Amazon.
>
> *...Science and technology must play a key role in sustainable development of the Amazon, considering the pressing necessity of new knowledge to fully develop the productive chains, starting with biodiversity and for valorising environmental services of ecosystems. It has thus become vital to develop a real scientific and technological revolution for the Amazon, a revolution held as the central and strategic priority of the regional development policy and that may possibly represent the greatest challenge to be faced by the Brazilian scientific community for the next thirty years....*
>
> *Technological capacity building has proved to be a fundamental tool to maintain the emerging economies of sizeable developing countries such as China, India and Brazil. Over the last fifty years, Brazil has been capable of creating islands of excellence in science and technology, which are more similar to those of developed countries than those of lower or middle income. However, historical regional inequalities, especially those in education, have created impediments drastically limiting intensive use of science and technology for the economies and social development of the poorer, less favoured regions, including the Amazon and the Brazilian Northeast....*
>
> *A new vision of science and technology is imperative. Among other general conditions, such as the improvement of basic education, it is essential to create a network of new institutions for higher learning, post-graduation, basic research and advanced technology with specific focus on both the forest and the aquatic resources. These institutions should be created so as to radically decentralize science and technology throughout the vast Amazon, maximizing the diversity and the potential of its sub-regions. Such an innovative network of science and technology should include five or six new technological institutions, grouping together from 500 to 600 faculty, researchers, engineers and technicians in each one, thereby multiplying the number of active researchers in the Amazon by three or four. In addition, these institutions – connected to a network of associated laboratories reaching every distant corner of the Amazon and interconnected by cutting-edge information technology – would serve as regional poles of this new technological development model....*
>
> *What the Amazon needs is many of these Amazon Technological Institutes to seed an innovative industrial model for that region. These institutes should be involved with the development and value aggregation in the entire productive chain of dozens of products from the Amazon, from bioprospecting, product development to commercialization and*

> *global marketing. Although it may seem a simplistic recipe for regional development, no tropical country has ever adopted it on a large scale. Cutting-edge technology would make it possible for some institutions to develop sophisticated research in biotechnology and nanoscience applied to biomimicry, that is, learning about the way complex biological systems find answers on a nanomolecular scale, to be reproduced in practical applications, a new scientific area to be explored for the tropical ecosystems.*
>
> Source: Nobre (2008)

4.6 Forest conservation

This chapter has looked at how commodities whose production is currently leading to deforestation can be produced in a more sustainable way. Putting a value on carbon has been identified as an essential incentive. Valuing the forest carbon externality (discussed in more detail in section 4.7) will also create further demand for activity which has the primary objective of preserving forests, in addition to the existing objectives of preserving landscape beauty, protecting biodiversity and promoting tourism.

This section examines two policy options for achieving this objective and assesses them in the light of the same aspects of sustainability as the production methods discussed previously.

4.6.1 Protected areas

Protected areas are likely to have a significant role in preserving global forests, but their design and management require the full participation of affected communities.

In theory, this traditional conservation model leaves forests almost entirely intact and is the most effective way to conserve forest carbon and the biodiversity and other ecosystem services they provide. Many protected areas also incorporate peripheral zones where managed economic activity can take place.

Evidence of their social impacts is mixed.[27] They can generate additional income from tourism, create employment in the form of park rangers and improve local environmental services such as water. However the potential for high levels of carbon finance from protected areas has prompted fears of a militaristic approach to mass forest management for carbon, potentially increasing the marginalisation of vulnerable populations.[28] Stringent restriction of human activity within some protected area boundaries can cause harm when communities are displaced or lose access to forest products they depend on or to land which has cultural/social value.[29] Income from tourism often does not reach the poorest and employment generated within protected areas is generally less than for other land uses.[30]

Overall, protected areas have a significant role in preserving global forests, but their design and management require the full participation of affected communities, and the challenges of ensuring sustainable livelihoods in and around parks should not be under-

---

27 Scherl et al (2004)
28 Griffiths (2007)
29 Smith and Scherr (2003)
30 Peskett et al (2008)

estimated. Box 4.4 describes a protected area programme in Indonesia and the methods it has established for resolving conflicts over land.

> **Box 4.4: Burung Indonesia – rural nature conservation agreements and participatory boundary demarcation**
>
> The Review Team met representatives of Burung Indonesia (BirdLife Indonesia), which works to:
>
> - provide support for improved planning and management of important sites, species and habitats;
> - introduce and advocate new ideas for integrating biodiversity conservation into planning and policy, for example through collaborative management and sustainable use of natural resources;
> - stimulate greater public interest in birds and biodiversity conservation;
> - develop improved management capacity;
> - provide information on biodiversity and protected areas to planners, policy-makers and other interest groups.
>
> Burung Indonesia has developed a participatory process for villages surrounding the Manupeu-Tanadaru National Park on Sumba Island, Karakelang Wildlife Sanctuary on Karakelang Island and Sahendaruman Protection Forest on Sangihe Island.
>
> The Rural Nature Conservation Agreement (RNCA) is a participatory agreement between government and communities within or adjacent to protected forests. The aim is to support the conservation and sustainable management of forest areas based upon traditional rules and norms of resource use. All stakeholders are brought into the process to reach agreements that are accepted by all involved. The final document typically consists of several agreements on how to solve problems such as addressing forest fires, protecting forest resources and the use of water sources such as springs.
>
> After the RNCA is reached, the first step is to implement the agreement through participatory boundary demarcation. The process results in agreement over the protected area boundary. In the Manupeu-Tanadaru National Park, Sumba, 270,860km of boundary has been agreed through facilitating the process with 18 villages.

4.6.2 Payment for ecosystem services

Payments for ecosystem services could provide a complementary income stream to that generated by some of the sustainable production methods described above.

A payment for ecosystem (or environmental) services (PES) is a transaction in which units of environmental service (ES), or a form of land use likely to secure that service, is bought by at least one ES buyer from a minimum of one ES provider if and only if the provider continues to supply that service (conditionality).[31]

---

31 www.cifor.cgiar.org/pes/_ref/about.index.htm

Payments can be for carbon and/or non-carbon ecosystem services. There are several large-scale PES schemes already in operation. Costa Rica hosts a national programme which rewards forest landholders for carbon sequestration, watershed protection, biodiversity conservation and the preservation of landscape beauty. Participants receive around $45 a hectare. The system is managed by FONAFIFO, a government agency. China's sloping land programmes pays farmers to replant trees on sloping land to prevent sedimentation, a potential; cause of flooding in the Yangtze.[32]

Chomitz describes the challenges of meeting equity and efficiency criteria through such schemes. Efficiency requires that only the most at risk forest is targeted, but the political viability of the scheme may depend on benefits being spread more widely. Logistical and administrative challenges are also considerable, and transaction costs in managing contracts can be high. Small landholders are underrepresented in the Costa Rica programme because it is cheaper to enrol large properties.[33] The establishment of PES schemes can help strengthen local institutions, but increasing land value can also exacerbate conflict and elite capture in less stable areas.[34]

The sustainability of pure PES schemes (and a large expansion of the protected area model) is questionable given their reliance on external finance, leading to concerns about the emergence of 'carbon dependency'. Leaving rainforests intact generates considerably less meaningful employment, for example, than other land uses. PES may be most appropriate as a complement to some of the production methods described above that extract value from land in a sustainable way.

4.7 Key levers for shifting to more sustainable production

Three key levers can help make the shift from deforestation to more sustainable policies and practices: valuing carbon and other services that forests provide; shifting policy incentives to more sustainable and efficient production practices; and using demand-side measures to support sustainable production.

These levers can operate at different levels – globally, nationally and locally. Putting a value on carbon is principally the responsibility of the international community as a key element of a new deal on climate change. The governance reforms required to shift policy incentives towards sustainable production will in large part be the responsibility of sovereign forest nations. Consumer countries, regional blocs and international agreements have a further role in putting in place demand-side measures to assist the shift. This section describes the need for, and operation of, these types of lever in more detail.

---

32 Chomitz (2006)
33 Zbinden and Lee (2004)
34 Peskett et al (2008)

4.7.1 Valuing carbon and other ecosystem services

The ecosystem services provided by forests, including forest carbon storage and sequestration, need to be valued.

We saw in Chapter 3 how it is often currently more lucrative to deforest and sell the resulting timber and agricultural produce than to leave forests standing. This is not because forests have no value in themselves, but because the costs of the deforestation are not reflected in the price of the timber or agricultural produce. The costs of deforestation can therefore be described as externalities (see Figure 4.3). This constitutes a market failure: the market will supply more timber and agricultural produce from deforested land than is efficient. If the costs of deforestation were factored into the price of products, their production would tend to shift to other land where they could be grown without deforestation.

Figure 4.3: Deforestation externalities: the current and true prices of products from deforested land

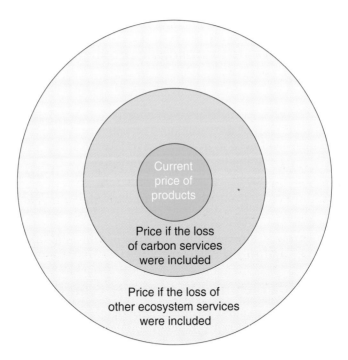

What is the magnitude of this externality? Forests provide many 'ecosystem services'. The storage and sequestration of carbon is just one of many. It has been estimated that the cost of forest ecosystems currently lost in just one year amounts to €1.35-3.1 trillion,[35] which is a huge figure and very much larger than profits to be made from the deforested land. There are a number of ways in which such externalities could be tackled, including:

---

35 Braat and Ten Brink (2008) Net present value over 50 years. The lower figure uses a 4 per cent discount rate; the higher figure uses a 1 per cent discount rate.

- *regulation* – a ban on deforestation or a ban on growing, selling or purchasing of products that have been produced from deforested land;
- *tax* – a tax on deforestation or on the growing, selling or purchasing of products that have been produced from deforested land;
- *cap and trade* – focusing on forest carbon services, under which forest owners could be given allowances to emit only up to a limited amount of carbon through deforestation and degradation.

All of these options could form effective policies for reducing the loss of forest carbon and other ecosystem services. But they all require that forest nations have the will and capacity to adopt and implement them. This is currently not the case, not least because the externalities are global externalities (ie, the whole world's population suffers from the loss of forests) whereas the profits to be made from deforestation accrue to individuals and government within the forest nation in question.

So, in order to pursue such policies, the governments of forest nations need first to be incentivised by the international community to bear down on deforestation. This is the central issue tackled by this Review. The options for addressing the externality of forest carbon will be analysed in more detail in Chapter 6.

4.7.2   Policy incentives for sustainable production practices

The policy and regulatory environment in forest nations needs to provide the right incentives to producers to make the shift to sustainable production.

Chapter 3 outlined the ways in which the governance framework in forest nations contributes to continued high rates of deforestation. The policy and regulatory environment in many forest nations will need to be reformed to take advantage of an international system of financial incentive and in recognition of the forest carbon externality. Appropriate laws, policies and programmes will be required to channel the financial incentive offered into the kind of sustainable production methods described above. Successful implementation of these policies is also likely to require a major capacity building effort for state and non-state institutions and actors. Chapter 12 looks in detail at the options forest nations have for accessing the global financial incentives for making the shift to sustainable production and preserving forests.

4.7.3   Demand-side measures in consumer countries

Demand-side policies in consumer countries, including preferential procurement of certified products as well as effective biofuels sustainability criteria that include the indirect effects of land use change, can also support the shift to sustainable practices.

A forest carbon value channelled through an appropriate national governance framework will provide an economic incentive for a shift towards more sustainable production methods. Demand-side measures in consumer countries (developed countries and emerging economies such as India and China) can also have a significant role in incentivising the shift to sustainable production, particularly in the short to medium term as new climate change mechanisms bed down.

Demand-side measures can help drive policy change, promote international cooperation on research and technology transfer, promote co-benefits, stimulate markets, and establish internationally agreed standards on what constitutes sustainability. This section looks at the role of legality assurance, sustainability standards and certification; public awareness; and the potential for reducing demand in some areas.

Legality assurance means developing systems that can ensure timber comes from legal sources. Where governance is weak, legality assurance is a more feasible way of improving the quality of forest management than certification. States would find it an easier option to deliver in the short term, as well as it acting as a stepping stone to sustainable production in the longer term.[36] Figure 4.4 describes the main differences between verification and certification.

A major legality assurance initiative underway is the EU's Forest Law Enforcement, Governance and Trade (FLEGT) Action Plan. This blends measures in producer and consumer countries to facilitate trade in legal timber and eliminate illegal timber from trade with the EU. The Action Plan sets out a range of measures including support to timber producing countries; activities to promote trade in legal timber; public procurement policies; and support for private-sector initiatives to promote corporate social responsibility. Voluntary Partnership Agreements (VPAs) commit both parties to develop systems for licensing legally produced timber from FLEGT partner countries and ensuring that only this timber is then allowed into the single EU market.

FLEGT explicitly recognises that some partner countries will require significant support to meet the requirements of the VPA, and that this technically and politically complex process may take some years of consistent and committed investments of time and technical expertise.[37] But engagement by rainforest nations in legality assurance schemes will be crucial to reducing deforestation, and the capacity they build in processes such as establishing legal clarity, broad stakeholder consultation and independent monitoring will be highly relevant when it comes to implementing wider measures. Rapidly growing economies such as China and India are an increasing source of demand for timber (often for processing and export to EU markets) and their active promotion of legality assurance would be a major contribution to international efforts to reduce deforestation.

Figure 4.4: Differences between verification and certification

Certification	Verification
Voluntary private sector initiative	Statutory government initiative
Global in scope and practice, but most advanced in the north	Global in scope, but in practice focuses on high risk countries
Aims at broad concept of sustainable forest management	Aims to achieve legal compliance with overall improved forest governance

---

36 www.verifor.org
37 Saunders et al (2008)

Sustainability certification aims to promote sustainable forest management, and add value for those who practise it. Preferential procurement of certified products from major importers can work with a carbon price to drive the adoption of sustainable production methods. Certification through standards such as Forest Stewardship Council (FSC) and national standards such as Brazil's CERFLOR promotes broader high standards of environmentally and socially sustainable management than basic legality verification. Box 4.5 describes the operations of one FSC certified logging company in Brazil.

Standards are also being developed to cover agricultural products, given increasing awareness of the GHG emissions from land use change. Sustainabilty criteria for biofuels used by suppliers to meet their targets under the Renewable Fuels and Fuel Quality Directives are currently being agreed by the EU. To be effective, it is essential that these criteria take into account indirect land use change impacts, ie the potential displacement of agriculture onto forest lands caused by increased production of biofuels.

Governments have an important role in driving markets for legal and sustainable timber, as recognised in the FLEGT Action Plan. In the UK, consumption of timber by central government is thought to account for 10 per cent of the market, rising to 40 per cent when wider public-sector purchasing is taken into account. But the influence of responsible purchasing by governments can be even wider. The private sector frequently looks to government to lead by example, adopting similar responsible purchasing policies as part of its broader Corporate Social Responsibility commitments. The timber trade in the UK has changed its practices significantly to meet the requirements of the central government timber policy; other consumers have benefited by proxy from this demand, and the market share of verified legal and sustainable timber has increased.

Forest certification has proved challenging in natural tropical rainforests (rather than plantations) and in situations of weak forest governance, particularly in parts of Africa.[38] Furthermore, despite the development of standards designed for small-scale producers, small and community-based enterprises have found certification prohibitively expensive and complicated. Linking certification, of agricultural products as well as forests, to international carbon finance may make it more viable for a wider range of producers.

Box 4.5: Cikel Brasil Verde Madeiras Ltda, Rio Capim Farm

The Review Team visited the FSC certified Cikel timber company in Brazil. Cikel produces a range of wooden flooring products. Eighty-five per cent of their products is exported, since there is low domestic demand for FSC products

Sustainable forest management for Cikel means:

- it is audited annually to retain certification (requiring multi-annual visits for the first 3 years);
- cut timber is tagged for traceability;
- extraction damage assessments, forest inventories and biodiversity inventories are carried out in the six-month rainy season during which logging ceases;

---

38 www.verifor.org

- it practises silviculture on a 35-year cycle, so 1/35th of its forest area is logged each year (in Rio Capim it is 4000ha) with around three to five trees taken from each ha producing 13-20m^3;
- it also reforests pasture (6000ha at Rio Capim) using native species such as mahogany which can be harvested after seven years and helps subsidise RIL of natural forest.

Acquiring certification from the FSC incurred up-front and additional running costs for Cikel, but it has brought with it the following advantages:

- fewer accidents;
- access to more profitable markets;
- more efficient process;
- a 75 per cent reduction of illegal logging;
- fewer invasions (land grabs).

Cikel is working with research institutes, communities and NGOs to increase the success of SFM by:

- working with the Tropical Forest Foundation to provide training for its forest technicians and help aid domestic education about SFM;
- linking with a university to investigate the feasibility of up to eight species plantations;
- supporting a successful study showing that removal of residuals (wood left after logging) increased regeneration and biodiversity and is profitable as a source for a charcoal industry that runs all year round on its site in Rio Capim;
- providing a school for the community.

Public awareness about the environmental impacts of producing wood and agricultural products is low. The role of the public in putting pressure on major procurers to insist on sustainable methods is an important one. In early 2008, for example, Staples cancelled contracts with Asia Pulp and Paper, one of the world's largest paper companies, because of what Staples called "their clear lack of progress in improving their environmental performance" in particular in relation to forest clearance in Indonesia.[39] Investment in public awareness campaigns and labelling relating to standards and certification could therefore add further pressure to make the shift to sustainability happen.

It is largely undesirable and impracticable to attempt to cut demand for agricultural and forest products, but the Review has identified two areas where *demand reduction measures* may be possible:

- Gallagher[40] proposes a slowing in the rate of biofuels expansion and a revision of the trajectories implied by current targets for their use in transport fuels. Biofuels targets (such as the EU Renewable Energy Directive) should be kept under review and modified as further evidence emerges of what constitutes a sustainable level of demand.

---

39 Bloomberg News (2008)
40 Gallagher (2008)

- Demand for fuelwood could be reduced by up to 70 per cent through the use of efficient woodburners.[41]

4.8 Conclusion

Increasing demand for commodities from a finite amount of land means that a step-change in the way agricultural and forest products are produced is essential. Shifting global production onto a more sustainable footing not only means reducing carbon emissions from land use, but could also make a major contribution to poverty reduction and the protection of non-carbon ecosystem services such as biodiversity and water systems.

International and national actors have a role to play in making this happen. The international community needs to put in place a system of financial incentives that reflects the global public good provided by the carbon storage and sequestration of forests. At the same time, forest nations will need to make the necessary governance reforms to ensure that these incentives have an impact on the ground, while consumers and consumer-country governments can aid the shift through preferences and demand-side policies.

The following chapter examines the cost of valuing carbon in more detail to understand the financial scale of the international challenge to shift to more sustainable production. Parts II, III and IV of this Review then go on to set out the principles of an international climate change framework that aims to meet the costs of mitigation.

---

41 www.hedon.org

5. The costs of mitigation

Key messages

The forest sector has significant potential for low-cost abatement. Realising this potential will incur opportunity costs (foregone profits from timber and agricultural commodity sales) as well as forest protection policy and administration costs. Some of these latter costs will need to be borne up front, while others will be ongoing.

Where the costs to a forest nation of conserving its forests outweigh the benefits, the international community should support forest nations in covering the shortfall given that the world as a whole benefits from reduced climate change damages.

This Review estimates that the finance required to halve emissions from the forest sector to 2030 could be between $17-33 billion per year if forests are included in global carbon trading. These results are based on various estimates from the literature and from modelling commissioned by this Review.

Risk modelling commissioned for this Review provides new evidence of the benefits of taking firm action to reduce forest emissions. Reducing deforestation rates significantly will require substantial finance. Nonetheless, the net benefits of halving deforestation could amount to $3.7 trillion over the long term (net present value). This is based on the global savings from reduced climate change minus the costs involved. The benefits would be even greater if deeper cuts to forest emissions were made or the preservation of other ecosystem services was taken into account.

5.1 Introduction

The previous chapter described how, in order to reduce (mitigate) forest emissions, forest nations will need to implement policies to bear down on deforestation. It also described how emissions from deforestation are part of the larger global challenge of reducing greenhouse gas emissions, from which the international community as a whole will benefit. This places a responsibility on the international community at large to support forest nations with finance to fund and implement mitigation policies.

This chapter looks at the different costs and transfers involved in mitigating forest carbon emissions through incentivising and bringing about more sustainable timber and agricultural production. It sets out the results of modelling commissioned by this Review to quantify the costs and also the benefits of taking action to lower forest emissions.

5.2 Up-front and ongoing mitigation costs

The forest sector has significant potential for low-cost abatement. Realising this potential will incur opportunity costs (foregone profits from timber and agricultural commodity sales) as well as forest protection policy and administration costs. Some of these latter costs will need to be borne up front, while others will be ongoing.

Where the costs to a forest nation of conserving its forests outweigh the benefits, the international community should support forest nations in covering the shortfall given that the world as a whole benefits from reduced climate change damages.

The mitigation costs of reducing forest emissions can first be sub-divided into two categories relating to the timeframe in which they will need to be incurred (see Figure 5.1 below):

a) **Up-front capacity-building costs.** These include the cost of building measuring and monitoring capacity so that an emissions reduction (or increase) can be accurately recorded, not least to permit a forest nation to claim a corresponding quantity of forest credits. Chapter 10 estimates that these costs will be around $50 million in the first year for a total of 25 forest nations (the running costs are given in Section 5.3 below). This cost category also covers the building up of governance capacity to enable forest nations to adopt and implement effective policies to reduce forest emissions. In Chapter 13 we report on work commissioned for this Review, which estimates these costs at up to $4 billion over five years for a total of 40 forest nations.

b) **Ongoing emissions reduction costs.** These cover the income foregone from avoided deforestation (opportunity costs) and the costs involved in adopting and implementing forest emissions reduction policies, including ongoing monitoring costs (forest protection costs). Both of these categories are further described below.

Figure 5.1: Mitigation cost categories for reducing forest emissions

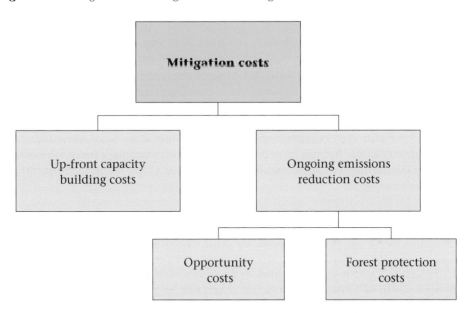

Forest nations will act to reduce forest emissions if they perceive that the benefits to them of acting are greater than the costs. The different types of cost that will need to be incurred have already been set out. The benefits for the forest nation will include receipt of payments for resulting forest credits as well as the further benefits resulting from maintaining forest ecosystem services, such as less flooding or droughts. If the costs to the forest nation are larger than the benefits, then the international community should support forest nations in covering the shortfall given that the world as a whole benefits from reduced climate change damages from forest emissions abatement.

5.3   Ongoing forest emissions reduction costs

As can be seen from Figure 5.1 above, ongoing emissions reduction costs are made up of two different components:

a) **Opportunity costs.** The opportunity costs of reduced forest emissions represent, as their name suggests, the costs of lost profit opportunities from not logging or converting forest land. The costs represent the profits that could have been made from sales of timber or of agricultural produce grown on the converted land. In Section 5.4 we report on work commissioned by this Review to provide an estimate of these costs.

b) **Forest protection costs.** These are the costs of adopting, implementing and administering policies to reduce forest emissions. Many different policies could be pursued to protect forest carbon and other ecosystem services, including designation and enforcement of protected areas; taxation of forest land clearance; restricting road building into forests; and agricultural zoning. Chapter 4 provides more details of the options that are available to forest nations for building up a portfolio of emissions reduction policies. The cost of monitoring emissions over time can also be included in this forest protection costs category. Chapter 10 reports on work commissioned for this Review, which estimates that for 25 countries the total recurring monitoring cost could be between $7-17 million per year.

The costs involved in administering forest protection policies (sometimes referred to as transaction costs) are substantial in their own right and should also be included within the forest protection costs category. A background paper commissioned for this Review estimates that the global administration, or transaction, costs involved in halving deforestation through the use of payments to forest landholders would amount to $233-$500 million per year, based on experience with payment for environmental service (PES) schemes in Latin America.[1] And an existing study examined the transaction costs involved in 11 moderately large forest carbon projects.[2] The mean transaction cost for producing the voluntary carbon credits generated was $0.38/tCO$_2$. The larger projects benefited from lower costs due to economies of scale. This study also showed that most types of forest project transaction cost (except for monitoring and verification) were on average lower than for other types of carbon mitigation project.

The forest protection costs for reducing forest emissions in any given nation will vary depending on the portfolio of policies that it chooses to pursue. For this reason, we do not attempt to give a cost estimate for global forest protection policies in this Review, although they should be borne in mind. Some protection policies (for example, a moratorium on building roads into forests) will be cheaper to finance than others. Each different potential policy will have wider implications for the forest nation in question and it is for that nation to determine which policies best suit its particular circumstances.

5.4 Estimating the opportunity costs of avoided deforestation

The opportunity cost of avoided deforestation is easier to estimate than forest protection costs. Opportunity cost modelling assumes that to avoid deforestation, the forest landholder would need to receive a similar payment to that obtainable from deforesting and selling off the timber and agricultural products from the converted land. As the sale price and costs of production vary between different commodities and the regions in which they are produced (see Chapter 3), so too will the cost of the foregone opportunity to profit from their production. This is illustrated by a marginal abatement cost curve (MACC) for reduced forest emissions, which shows how cheap forest abatement opportunities become exhausted as greater emissions reductions are made (see Box 5.1).

---

1 Grieg-Gran (2008)
2 Antinori and Sathaye (2007)

Box 5.1: Marginal abatement cost curves (MACCs) for reduced forest emissions

A marginal abatement cost curve traces out, for a given point in time, the cost of the last unit of abatement (see Figure A). It is derived by plotting, for a given year, the amount of abatement occurring in forest nations for any given carbon price. The MACC can be thought of as an abatement supply curve that displays the level of abatement achieved in the forest sector for any given carbon price.

Figure A: Marginal abatement cost curve for global forest abatement in a given year
Potential forest emissions abatement is initially relatively cheap.

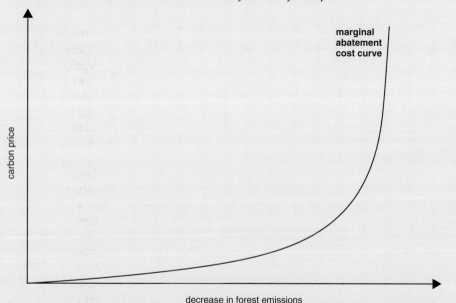

This abatement consists of avoiding forest conversion to the lowest land-use returns, such as subsistence agriculture in Africa and small-scale beef farming in South America. This lowest cost abatement is represented by the area under the left-hand side of the curve in Figure A. It should be noted, however, that where the opportunity cost is low the forest protection costs could be disproportionally high (as there might be particularly low governance capacity).

As one follows the curve to the top right-hand side, it becomes increasingly expensive to supply an extra unit of forest abatement. It is here that the highest land use returns foregone (such as large-scale oil palm plantations in South East Asia) have to be compensated.

This Review commissioned as a background paper[3] an update of the opportunity cost estimate for avoided deforestation produced for the Stern Review.[4] This work takes a bottom-up approach, using the following steps:

a) establish the area of forest required to be maintained;
b) assess the likely alternative use of the forest land that is to be maintained;
c) consider current (and where possible forecasts of) commodity prices (see Chapter 3 for a discussion of commodity prices);
d) estimate the net present value of future income streams that could be made from converting forest land to its different uses (using a discount rate);
e) multiply the net present values by the area of forest to be protected from conversion to each use.

The land-use returns that are foregone if deforestation is halted completely were calculated for eight countries, which together represent 46 per cent of global deforestation. It was assumed that the area in question is conserved over a 30-year period. For the most likely scenario (taking into account legal, practical and market constraints on logging), the opportunity cost for roughly halving global deforestation was estimated at around $7 billion per year. This estimate is higher than the $5 billion estimated for the Stern Review because of rises in agricultural commodity prices over the last two years. The commodities that give the highest land-use returns, such as oil palm, were particularly influential in inflating the cost.

This result is broadly consistent with other recent bottom-up opportunity cost calculations. In 2007, an estimate was prepared for the UNFCCC Secretariat of the opportunity cost of reducing forest emissions to zero by 2030.[5] It found that a minimum investment of $12.2 billion per year would be needed, and that an average carbon price of $2.8/t$CO_2$ would reduce forest emissions by 65 per cent.

Several major assumptions behind most opportunity cost modelling, including the modelling that produced the $7 billion figure, should be noted. Most importantly, the $7 billion represents the amount that authorities administering payments to landholders would incur if they could target them and pay different rates according to individual opportunity costs. In practice, such price discrimination is unlikely to be feasible (and could also be regarded as inequitable).

Other assumptions included in the modelling are that:

- the areas most at risk from deforestation can be accurately identified and targeted, ensuring 100 per cent additionality – this is unlikely in practice, which could mean that the $7 billion figure is an underestimate;
- leakage does not occur – this is unlikely to be achievable in practice, which could also mean that the $7 billion figure is an underestimate;
- landholders derive no benefit from standing forests, which is not the case in practice. For example, forest foodstuffs can be sustainably harvested and sold for profit or used for subsistence. This could mean that the $7 billion figure is an overestimate as the net financial benefit of converting the forest to an alternative use would be lower.

---

3   Grieg-Gran (2008)
4   Stern Review (2007)
5   Blaser et al (2008)

5.5 Estimating the costs of purchasing forest emissions abatement

This Review estimates that the finance required to halve emissions from the forest sector to 2030 could be between $17-33 billion per year if forests are included in global carbon trading. These results are based on various estimates from the literature and from modelling commissioned by this Review.

This Review recommends that in the transition to a comprehensive global cap and trade system, non-Annex I forest nations should have the right to generate and sell credits representing reductions in emissions (see Chapters 8 and 9). This section looks at the question of how much it could cost Annex I countries and companies to purchase forest emissions abatement from forest nations in the form of forest credits sold on the international credits market.

Some of the world's leading experts on mitigation cost modelling in the forest sector published a paper in 2008 setting out the results of three different global land use and management models.[6] These models were used to estimate the cost of purchasing credits from an open carbon market to realise part of the world's total forest abatement potential. The results of the modelling included a range for the average annual costs for halving global forest emissions between 2005 and 2030: $17.2-28.0 billion per year.

This Review commissioned runs from two of the three models featured in the paper (the IIASA model cluster and GCOMAP).[7] The model runs for this Review used updated input data and included some improvements to the models themselves. Further details of the modelling are contained in two background papers to this Review.[8] In broad terms, the modelling used the following steps to determine a series of MACCs:

a) future deforestation emissions are projected under a business as usual scenario, using various economic and forest biomass assumptions. This constitutes the baseline (where forest carbon is worth $0 per ton CO_2);

b) the costs of carbon storage resulting from avoided deforestation are calculated, using further land use and management simulations with carbon prices between $0 and $100 per ton CO_2;

c) carbon emissions are compared between the baseline scenario and those scenarios where avoided deforestation is compensated.

The models make a number of assumptions in addition to those set out in Section 5.4 above, including about future trends in population, technology and trade. Each model also makes different emissions projections (due to different assumptions regarding the number of hectares lost and the biomass content of those hectares of forest).

The MACCs derived from the commissioned modelling were used by this Review to estimate the cost of purchasing forest credits representing half of global forest abatement potential to 2030. The average annual cost from the IIASA cluster model MACCs was $22 billion per year. The figure from the GCOMAP MACCs was $33 billion per year.[9] When combined with the spread of $17.2-28.0 billion reported above, this provides a

---

6   Kindermann et al (2008)
7   The remaining model (GTM) was, incidentally, responsible for the lower end of the $17.2-28.0 billion range.
8   Sathaye et al (2008) and Gusti et al (2008)
9   The $22 billion and £33 billion figures translate into average carbon prices of $11 and $15/ tCO_2 respectively.

range of **between $17-33 billion per year** to 2030 to halve global forest emissions through carbon trading.

Unlike the $7 billion opportunity cost figure reported in Section 5.4, these modelling results cover not only the opportunity cost but also rent. The opportunity cost can be thought of as simply the cost of *supplying* forest abatement, which is represented by the dark green area below the MACC in Figure 5.2. As discussed above, opportunity cost modelling assumes that landholders are paid at different rates according to the opportunity cost for their particular plot of land. In addition, the estimates of $17-33 billion include the profits, or rent, received by forest credit sellers who supply their credits below the marginal cost of the last unit of abatement. In an open credit market, all credits would tend to be sold at the price where supply and demand for credits are equalised (price p* in Figure 5.2), despite the fact that the majority of credits would be cheaper than this to supply. The resulting rent that the forest credit suppliers would receive is shown in the lighter green shaded area in the figure. In the $22 billion and $33 billion figures reported above, rent constituted $9 billion and $18 billion respectively of the total amounts.

Figure 5.2: MACC highlighting forest abatement supply costs and rent

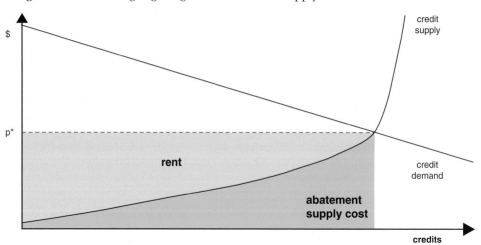

In practice, it may not be possible to purchase a sufficient number of credits via the open carbon market at its current stage of development (see Chapter 11). If some of the credits are purchased with non-carbon market finance, then there are arguments either way as to whether full rent should be paid. On the one hand, a strong incentive is essential for nations to act to reduce their forest emissions. Many forest nations rank among the world's poorest and any rent they did receive might be comparable to development aid. Rent will tend to be higher in poorer countries where the opportunity costs are lower. Some would argue that all countries and actors should receive an equitable price for their services, ie the carbon emissions they reduce, regardless of whether they are poorer or richer. Furthermore, countries where opportunity costs are the lowest are often those where the policy, administration and monitoring challenges are greatest.

On the other hand, it could also be argued that poorer countries would benefit from an institution that buys credits at a guaranteed minimum purchase price (see Chapter 11). This could result in lower rents but greater price certainty. Others may argue that it is less acceptable to pay rent from public funds constituting taxpayer money, particularly where recipient nations are relatively wealthy and receive significant local benefits from forest conservation. Furthermore, the fact that 84 per cent of the world's forest land is publicly owned[10] may also have a bearing on this issue. Ultimately, the level of financial transfers to forest nations should be based upon an assessment of the costs and benefits to these countries of reducing forest emissions, and the extra incentive that the international community should pay to cover any shortfall.

5.6 The benefits of taking action to reduce forest emissions

Risk modelling commissioned for this Review provides new evidence of the benefits of taking firm action to reduce forest emissions. Reducing deforestation rates significantly will require substantial finance. Nonetheless, the net benefits of halving deforestation could amount to $3.7 trillion over the long term (net present value). This is based on the global savings from reduced climate change minus the costs involved. The benefits would be even greater if deeper cuts to forest emissions were made or the preservation of other ecosystem services was taken into account.

This Review commissioned a background paper to estimate the contribution of forest emissions to climate change impacts and an assessment of the costs and benefits of taking action to reduce those emissions.[11] In Chapter 2 we saw that the damage costs of continued climate change from the impact of business as usual (BAU) forest emissions from 2010 to 2200 were estimated at around $12 trillion (mean net present value). Using similar modelling,[12] and with Houghton's deforestation projections,[13] a 50 per cent reduction in forest emissions from BAU was found to reduce the damage costs from climate change impacts by $5.3 trillion, calculated as a mean net present value (see Figure 5.3a).[14]

The model was then used to calculate the costs of purchasing forest credits from halving forest emissions from 2010 to 2200, estimating unit costs from the Kindermann et al (2008) paper and converting them to an annual payment form. This gave a mean net present value of around $1.7 trillion (see Figure 5.3b).

---

10 FAO (2006)
11 Hope and Castilla-Rubio (2008)
12 See Chapter 2 for a description of the model's methodology.
13 Using these projections in the model, forest emissions drop to zero at 2100 due to the lack of forest area remaining to be converted.
14 A pure time preference rate of 0-2 per cent was used (most likely value of 1) with an equity weight of 0.5-2 (most likely value of 1). Using these values gives discount rate ranges for the various regions modelled of 1-5 per cent.

Comparing the reduction in damage costs on the one hand and the financial costs of reducing emissions on the other gives a **mean net benefit of around $3.7 trillion** for halving forest emissions between 2010 and 2200 (see Figure 5.3c). This figure is based on various assumptions underpinning the model, which are described in the background paper. It should be regarded as simply indicative of the likely scale of the net benefits of reduced deforestation.

Given that the probability distribution of the net benefit has a long upper tail (Figure 5.3c) due to the high costs associated with the more extreme scenarios of climate change impacts, the figure could be considerably higher. Nor does the $3.7 trillion figure include the benefits of preserving other forest ecosystem services. The damage costs of all forest ecosystem services (including climate change regulation) lost in just one year have been estimated elsewhere at €1.35-3.1 trillion.[15]

Figure 5.3: Global benefits of reducing forest emissions (mean net present value showing probability distribution)

(a) Reduction in climate change damage costs from halving forest emissions

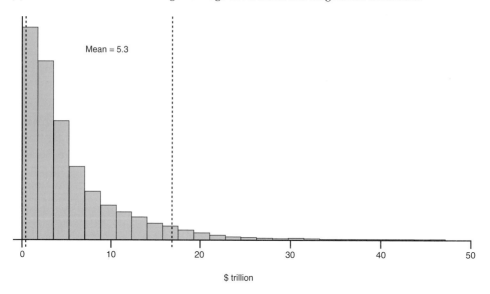

---

15 Braat and Ten Brink (2008)

5. The costs of mitigation

(b) Financial cost of halving forest emissions

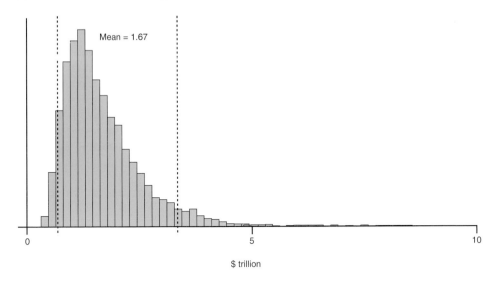

(c) Net benefit of halving forest emissions

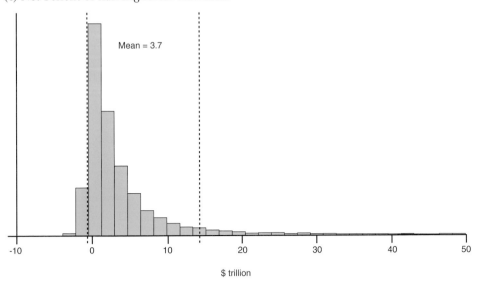

Note: The grey vertical lines represent confidence intervals at 5 and 95 per cent on the probability distribution.

The model was also used to investigate the costs and benefits of undertaking more ambitious cuts in forest emissions. The results show that the more ambitious the cuts, the greater the net benefit. For example, a 90 per cent reduction in forest emissions led to a global mean net benefit of $6.3 trillion.

Using different projected forest emissions input data gives net benefit results of a similar scale. This was shown from further model runs using the IPCC's SRES A2 projections of forest emissions. The results of these further runs for halving deforestation were as follows: mean reduction in damage costs of $5.6 trillion; mean cost of halving emissions, $1.2 trillion; and a mean net benefit of $4.4 trillion. This net benefit is around 20 per cent higher than the net benefit using the Houghton emissions projections. These results show that the mean net benefit results are quite insensitive to the global greenhouse gas stabilisation trajectory upon which they are superimposed (see Chapter 2 for more details).

5.7   Conclusion

Forest emissions are part of the larger global challenge of climate change. This places a responsibility on industrialised nations to support developing countries in reducing greenhouse gas emissions within their borders. This Review estimates that the finance required to halve emissions from the forest sector to 2030 could be around $17-33 billion per year if forest credits are included in global carbon trading. This represents the cost of buying half of all global forest emissions abatement potential in the open carbon market.

Although this cost is large, it is substantially outweighed by the benefits of reducing the global damage costs from forest emissions-induced climate change. On the basis of modelling commissioned by the Review, we estimate a mean global net benefit from halving forest emissions of around $3.7 trillion (net present value). And the more ambitious the cuts the greater the net benefit becomes, rising to around $6.3 trillion when forest emissions are reduced by 90 per cent.

The following chapters consider the most appropriate long, medium and short-term funding sources and systems for bringing about the deep cuts in forest emissions that will be necessary to give the world a realistic chance of preventing an average temperature rise of more than 2°C and avoiding the worst impacts of climate change.

Part II:
Forests and the international climate change framework: the long-term goal

International finance will be key to tackling global deforestation. Part I set out a vision of sustainable production that leads to reduced forest carbon emissions, better livelihoods for forest communities – some of whom are the poorest in the world – and protection of biodiversity and other ecosystem services. It also examined the potential financial costs, particularly in the short to medium term, of shifting from inefficient, unsustainable deforestation to more sustainable agricultural and timber production.

Part II examines the long-term international framework required to reduce emissions from deforestation and forest degradation and to provide incentives for afforestation, reforestation and restoration in the most effective, efficient and equitable manner.

Chapter 6 sets out the economic rationale for international collective action and provides a long-term framework for financing reductions in forest carbon emissions. It concludes that, while various sources of funding should be used in parallel to finance forest emissions reductions, a global cap and trade system that fully includes the forest sector performs best against the criteria of effectiveness, efficiency and equity.

Chapter 7 examines the current international framework of carbon targets and trading under the United Nations Framework Convention on Climate Change. It looks at the advantages and shortcomings of existing approaches, and examines how far the international community is from a fully functioning global cap and trade system. The chapter concludes that, while cap and trade should be the long-term global goal, a period of transition in the short to medium term will be required.

6. A long-term framework for tackling climate change

Key messages

Forest carbon emissions need to be tackled as part of an overall approach to addressing climate change. Therefore the overall international climate change framework needs to be the starting point for developing forest solutions.

Any future international climate change framework should be based on three criteria: effectiveness, efficiency and equity. The framework should be effective to deliver the emissions reductions at the required scale; efficient to minimise the overall cost of achieving the emissions reductions; and equitable to ensure that the benefits of international action are distributed fairly.

To be effective, an international emissions reduction system needs to tackle three major challenges for all sectors. First, it should ensure that mitigation activities in one area do not lead to *leakage* of emissions elsewhere (eg industrial or forestry companies relocating). Second, reductions should be *additional* to what would have occurred in the absence of intervention. Third, the system needs to guard against the risk of *impermanence* to ensure that emission reductions are locked in over time.

A variety of systems exist for achieving reductions in deforestation as part of an overall global framework. Many of these systems are potentially valuable tools in tackling climate change and could act in parallel. The two principal options for valuing forest carbon in the long-term are taxation and cap and trade. A global cap and trade system performs best against the criteria of effectiveness, efficiency and equity.

Integrating forests within a global cap and trade system would create opportunities to tackle a large proportion of current CO_2 emissions while delivering substantial finance for forest conservation and sustainable forest management. Excluding the forest sector would impede the benefit of trading to maximise emissions reductions and minimise costs.

Including reduced emissions from deforestation and degradation (REDD) in a well-designed cap and trade system could reduce emissions from deforestation by up to 75 per cent in 2030. With the addition of afforestation, reforestation and restoration (ARR), this would make the forest sector carbon neutral.

Including forests in a global cap and trade system would mean that the cost of halving global carbon emissions from 1990 levels could be reduced by up to 50 per cent in 2030 and by up to 40 per cent in 2050.

> This could allow the international community to meet a more ambitious global stabilisation target. By 2050, CO_2 emissions could be reduced by an additional 10 per cent with the inclusion of the forest sector.
>
> Forest carbon finance could also make a significant impact on reducing poverty through increased financial flows to developing countries.

6.1 Overall framework for tackling climate change

Forest carbon emissions need to be tackled as part of an overall approach to addressing climate change. Therefore the overall international climate change framework needs to be the starting point for developing forest solutions.

Tackling forest emissions needs to be part of an overall approach for tackling climate change in all sectors. A central aim of the UNFCCC is to achieve stabilisation of greenhouse gas concentrations in the atmosphere at a level that would prevent dangerous anthropogenic interference with the climate system.[1] Preventing dangerous interference should allow ecosystems to adapt naturally, ensure continued food production, and allow sustainable economic development.

It is widely suggested that the increase in global temperature that is currently occurring should be stabilised at a maximum of 2°C over pre-industrial levels to minimise the risk of dangerous climate change.[2] IPCC has indicated that achieving a 2°C target would mean stabilising GHG concentrations in the atmosphere at around 445 to 490 ppm CO_2e or lower.[3] Against a background of rising population and increasing prosperity, this will require a reduction in annual global emissions of around 50 per cent or more from 1990 levels by 2050 (see Figure 6.1). Given that forestry, as defined by the IPCC, contributes around 17 per cent of current GHG emissions, net emissions from forests will need to be substantially reduced if this goal is to be achieved.

As discussed in Chapter 4, these essential but ambitious long-term emissions reductions will require a step-change in the way land is used, so that pressures to deforest become positive incentives to keep forests standing and managed in a sustainable way. This means addressing the global externality that currently exists for carbon and CO_2 emissions so that the social cost of releasing CO_2 into the atmosphere, as well as the social benefit of reducing these emissions, can be realised. Therefore an overall international framework will need to facilitate a behavioural step-change and bring a true value to natural standing forests. As Chapter 5 explains, significant funds will be required to implement this step-change and to address the opportunity costs (income foregone) on a global scale.

This chapter will examine systems that could be used on a global scale to tackle the externality of CO_2 emissions from all sectors effectively, efficiently and equitably. We will then examine the rationale for integrating forests into an overall climate change framework. Generating a carbon value will provide the incentives to overcome the barriers that

---

1 United Nations (1992) UNFCCC
2 EU target for emissions reduction
3 IPCC (2007) WG 3 Chapter 9

currently exist when tackling the drivers of deforestation. Although it will not be feasible to fully implement this long-term framework immediately, it will be essential to establish a goal in order to enable a smooth, measured and predictable transition.

Figure 6.1: Global emissions path required for stabilisation at 475ppm CO_2e overshooting to 500ppm

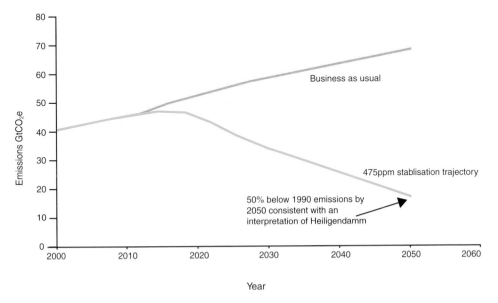

Note: *Heiligendamm refers to the G8 Summit in Heiligendamm , 6-8 June 2007. The current level of emissions together with the stabilisation trajectory have come from the SiMCaP model. Forecast business as usual emissions have come from three GHG models. Energy CO_2 from the POLES model, forestry emissions from the IIASA cluster model and non-CO_2 emissions from the IMAGE model.*

6.2 Criteria for a successful climate change framework

Any future international climate change framework should be based on three criteria: effectiveness, efficiency and equity. The framework should be effective to deliver the emissions reductions at the required scale; efficient to minimise the overall cost of achieving the emissions reductions; and equitable to ensure that the benefits of international action are distributed fairly.

Forest carbon emissions, along with emissions from other sources, are a global negative externality. The cost of each unit released into the atmosphere is not borne by the emitter. Instead the costs are imposed on the international community as a whole in the form of exposure to the damaging effects of climate change. Conversely there is no comprehensive system that rewards reduced emissions from deforestation, for example, even though it brings global benefits. Therefore, in order to incentivise forest nations to take action on deforestation, it is reasonable that any international framework to tackle climate change should internalise the emissions from forests. Bearing this in mind, there are three criteria that a successful international climate change framework should meet: effectiveness, efficiency and equity.[4] The following sections set out these three criteria in more detail.

6.2.1 Effectiveness

To be effective, an international emissions reduction system needs to tackle three major challenges for all sectors. First, it should ensure that mitigation activities in one area do not lead to leakage of emissions elsewhere (eg industrial or forestry companies relocating). Second, reductions should be additional to what would have occurred in the absence of intervention. Third, the system needs to guard against the risk of impermanence to ensure that emission reductions are locked in over time.

Given the potential consequences of dangerous climate change, it is essential that the system is effective in achieving reductions in emissions at the right scale. Whether payments are to be transferred under a trading system or imposed through taxation, any system would need to achieve sufficient scale to deal with the geographic spread and severity of emissions.

An international emissions reduction system for all sectors also needs to tackle three major challenges to be effective: leakage, additionality and permanence (see Box 6.1). A successful international framework needs to cover the sources of emissions comprehensively, regardless of the country or sector they originate from. Otherwise, there is the risk that emissions will migrate, or leak, to sources that lie outside the system. Leakage could occur across areas within a country or between countries. Therefore any system will need to have comprehensive participation by the major emitters at national and international levels to prevent leakage. To achieve this, a successful system will need to be comprehensive in terms of its coverage of sectors and politically acceptable to the main emitting countries.

The second requirement for ensuring effectiveness is that emissions reductions should be additional to any reduction that would have occurred in the absence of intervention. This means that any system needs to minimise non-additional reductions that are accidentally or unfairly credited.[5] If an abatement activity gains credits for reductions that would have occurred anyway, the environmental integrity of the credit awarded is undermined (see Box 6.1).

The third consideration for ensuring that the system is effective is to ensure that carbon reductions are locked in over time – permanence. Stabilising the carbon stock requires a permanent reduction in the flow of emissions into the atmosphere, which is an issue for all sectors. For example, the recent increase in gas prices relative to coal has led to an increase in emissions from the UK power sector. Similarly, economic conditions can affect deforestation rates. Furthermore, forests are vulnerable to disturbances, such as drought, fire or pests.

---

4   These criteria were also set out in Stern's 'Key elements of a global deal' (2008) when putting forward a set of proposals on global policy to provide economic incentives for addressing the impact of global warming.
5   Bloomgarden and Trexler (2008)

These can be naturally or human-induced and can result in the release of stored carbon (see Box 6.1).

> **Box 6.1: Effectiveness of an international climate change framework: leakage, additionality and permanence**
>
> An international emissions reduction system needs to tackle three major challenges for all sectors if it is to be effective: preventing leakage of carbon emissions to other locations; ensuring that emissions reductions are additional to reductions that would have occurred in any case; and the risk that reductions are not permanent.
>
> **Leakage**
> Leakage occurs when mitigation activities in one place cause an increase in emissions elsewhere. This is a challenge in a number of sectors, such as industry. Leakage may occur through two main channels. Direct leakage could occur if, for example, heavy industry with high CO_2 emissions decided to relocate to countries that were not participants in the system rather than bear the costs of reducing emissions. Indirect leakage could occur if, for example, reduced energy demand in Annex I countries with emissions reduction targets led to a reduction the price of energy in the region. The lower energy price could then stimulate demand in countries that do not have emissions restrictions.[6]
>
> The risk of leakage has been perceived as more severe for forests than for the transport or industry sectors. While some modelling has suggested that forests are more prone to leakage, real world studies have been less conclusive.[7] Leakage can occur with any mitigation activity. The perception that forests are different from other sectors may instead be based on the nature of the CDM as a project-based system. Using a national-level baseline or reference level and implementing the measuring and monitoring of forest emissions on a large geographical scale would significantly reduce the risks associated with project-level leakage, particularly within countries (see Chapter 9). International leakage is a challenge for all sectors and needs to be addressed with comprehensive coverage of an international emissions reduction system such as a global cap and trade system.
>
> **Additionality**
> A system for mitigating emissions should ensure that emissions reductions are additional to any reduction that would have occurred in the absence of intervention, ie under a business as usual scenario. Ensuring additionality creates a challenge when attempting to quantify effective reductions in carbon emissions, as the business as usual scenario that would have occurred without intervention is unobservable. If there are problems with accurately projecting business as usual, the effort devoted to reducing emissions could be overstated. This, in turn, could lead to the international community paying for 'hot air'. Lack of additionality is a challenge in all sectors and needs sufficient ambition in setting targets as well as robust measuring and monitoring.

---

6   Lejour and Manders (1999)
7   Schwarze, Niles and Olander (2002)

> **Permanence**
>
> Climate change results from changes in the concentration of CO_2 in the atmosphere. Stabilising this stock requires a permanent reduction in the flow of emissions into the atmosphere, which is an issue for all sectors. For example, the recent increase in gas prices relative to coal has led to an increase in emissions from the UK power sector.[8]
>
> A specific concern that has been cited for including the forest sector in a global scheme is that forests are more vulnerable than other sectors to natural disturbances, such as drought, fire or pests, which can cause the release of stored carbon into the atmosphere. This can be a particular problem for ARR projects. The slow and gradual uptake of CO_2 as new forests grow can be reversed relatively quickly through an environmental disturbance. Although this is a risk for local projects, it is likely to be less significant within the wider coverage that national-level accounting provides.
>
> It has also been argued that there is value in temporarily delaying the release of carbon into the atmosphere, particularly through reduced emissions from forests, as it reduces the amount of cumulative carbon stock in the atmosphere. This can reduce the impacts that are caused by climate change by reducing emissions in the short to medium term while new technologies and economic instruments are being put in place to reduce industrial emissions.[9] Taking long-term responsibility for the carbon stocks will ensure that the potential future release of carbon is accounted for. Nonetheless, permanence is an important challenge for forests as well as other sectors and is discussed in Part III of this Review.

6.2.2 Efficiency

The Stern Review estimated that the costs of achieving a global stabilisation target of 550ppm CO_2e by 2050, could be limited to, on average, around 1 per cent of global GDP per year.[10] However, as Stern has pointed out, this is unlikely to be realised unless the international system for achieving these reductions is efficient. Therefore to achieve a more ambitious stabilisation target of around 445 to 490ppm CO_2e efficiency will need to play a key role. Transaction costs – the costs incurred in the administration of the system – will form an important part in determining the efficiency of the system (they are discussed in more detail in Chapter 5). Increasing efficiency reduces the global cost of achieving emissions reductions which would also allow for more ambitious emissions reductions to be achieved.

One of the main requirements in economic theory for achieving efficiency is that the costs of abating the last unit of CO_2 are the same in every sector and every country. This is because the damage caused by each tonne of CO_2 is the same regardless of where it is emitted. If there were spatial or sectoral variations in costs, the total cost could be reduced by abating less where it is expensive and more where it is cheap.

---

8   DTI (2007)
9   Fearneside et al (2000)
10 Stern (2007)

The second requirement for a system to be efficient is that there should be abatement up until the point where the cost of abating the last unit of CO_2 exceeds the benefits of doing so. This means that – all other things being equal – it will be efficient to reduce the total amount of emissions in response to cheaper abatement options becoming available. There are sectors where the price signal alone does not lower costs efficiently, for example in expensive technologies which require learning and experience in order to induce cost-cutting innovation, or efficiency gains which are hard to generate through price incentives. For the most part, however, prices play a central role in guiding markets towards an efficient allocation of resources.

Given uncertainty over abatement costs in different places, and how they are likely to change over time, an efficient system should provide flexibility over where reductions occur.

6.2.3 Equity

While the efficiency criterion considers the cost to the world as a whole, the equity criterion assesses who pays and who benefits. Equally efficient mechanisms can have very different distributional outcomes. In theory, it is possible to consider the two separately, but in practice the choice of system is likely to have intra-national and international distributional consequences.

International action to reduce emissions will require the willing participation of a broad range of countries. This will happen only if the system is able to offer a distribution of costs and benefits that is perceived to be fair. This is particularly important in the forestry context, given that deforestation mainly occurs in developing countries, including many of the least developed countries. Given that there are many ways in which fairness could be defined, it is essential that the system can provide the flexibility for this to be negotiated.

As well as the distribution of effort between countries, it is also important to consider the distribution within countries. For example, distribution will be particularly importatant for the 90 per cent of people living on less than $1 per day who rely on forests to some extent for their livelihoods.[11] Consideration of the needs and circumstances of developing countries must therefore be provided within the system. Ensuring national sovereignty, specifically with regard to land-use decisions, while preventing misuse of the system will need to be key components. This will involve the provision of flexibility when delivering funding or setting emissions caps, to take into account a country's specific interests, profile and circumstances.

---

11 Scherr et al (2003)

6.3 Comparison of options for achieving global climate stabilisation

A variety of systems exists for achieving reductions in deforestation as part of an overall global framework. Many of these systems are potentially valuable tools in tackling climate change and could act in parallel. The two principle options for valuing forest carbon in the long term are taxation and cap and trade. A global cap and trade system performs best against the criteria of effectiveness, efficiency and equity.

In order to address the global externality of CO_2 emissions, carbon needs to be valued to represent the price (social cost) or penalty that would be paid by those who generate the CO_2 emissions.[12] A range of approaches to value carbon is provided by economic theory. These can be categorised under two broad systems: a price could be imposed through a carbon tax, or could arise from a trading system (see Figure 6.2).

Figure 6.2: Options for addressing the externality of CO_2 emissions from forests

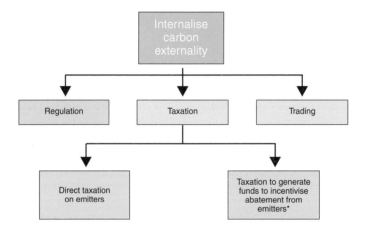

*This could include indirect taxation such as tariffs on the trade in timber and non-timber products

While not explicitly imposing a carbon price, environmental regulation could also be an option for addressing the externality of CO_2 emissions. An example of regulation on an international scale is the EU Forest Law Enforcement, Governance and Trade (FLEGT) Action Plan. Agreements are set up between the forest nation and the consumer countries to implement measures for trading in legal timber and for eliminating illegally logged timber from trade with the EU. Chapter 4 discusses the importance of the EU FLEGT in more detail. Both taxation and trading have the potential to generate public funds. For taxation, this occurs automatically as the contribution is imposed by government. With trading it occurs if purchasers are obliged to buy quotas from government through sales or auctions.[13] The important role of public funds (multilateral and bilateral funds) to

---

12 Nordhaus (2008)
13 Stern (2007)

generate up-front finance and supplement forest funding in the short and medium term, will be discussed in more detail in Chapters 11 and 13.

6.3.1 Carbon taxation

Carbon taxation, at a domestic level, is undoubtedly one tool with which policy makers will wish to tackle climate change. In its simplest state, taxation could address the externality by levying a uniform tax on all sources of CO_2 emissions at a rate equal to the damage caused by each tonne emitted.

Taxation avoids the problem of determining *additionality*. At the sub-national level, systems of positive incentives are faced with the challenge of determining what would have happened in the absence of incentives not to reduce emissions. This is problematic because the emitter would have the incentive to overestimate the emissions that would have been released in the absence of payments. The administering authority is then faced with the challenge of assessing those claims. Taxation avoids this difficulty because it is levied on behaviour that is potentially observable. As such, it avoids the need to determine what would have occurred in the absence of payment. However, in the case of afforestation where trees are planted, an assessment of whether they would have been planted without mitigation incentives would be necessary in order to generate credits. Moreover, taxes provide a disincentive to cut trees but not an incentive to monitor or declare. By contrast, credits provide an incentive to preserve trees and monitor and declare.

Domestic carbon taxation would overcome the problem of intra-country *leakage*, provided that it was levied on all sectors. Activities that led to emissions would face a carbon price regardless of where they occurred, so in theory there should be no risk of them being relocated. International leakage, however, could still occur if there are variations in the effective carbon price between regions. This variation could exist from a lack of incentive to participate in harmonisation, or an inability to enforce the taxes that they have. Either way, the risk would be that emissions were displaced rather than reduced.

The *permanence* of reductions in emissions from taxation is less certain. It would depend upon the durability of the mechanism. Taxation would provide emitters with the incentive not to emit only for as long as it was in operation. If the tax ceased to be in operation, then there would be nothing to prevent landowners from clearing any land that had previously been protected, other than technical or economic limits on the rates of deforestation. Harmonisation of domestic taxation would be viable only for as long as the accompanying transfers were in operation.

To be an efficient system, carbon taxation would need to address the externality on an international scale so that all emitters would face a uniform carbon price. However, decisions over taxation are made at a regional or national level. As a consequence, the levels of domestic carbon taxes would need to be harmonised internationally to ensure a uniform carbon price. For example, the carbon tax paid by road users in Italy would need to be consistent with the carbon tax faced by a steel factory in China or a soy farmer in Brazil. Such consistency would require coordination at an international level to eliminate major differences and to create requirements or minimum standards. This would be particularly challenging as it would involve political agreement and international consensus across all sectors and governments. These difficulties have been witnessed in the failure to agree a common carbon tax by European countries.[14]

In order to be equitable, a regime based on taxation would need to be able to allocate a greater share of the responsibility for reducing emissions to developed countries. This is clearly important on the grounds of fairness but would present the system with difficulties. Taxation may be compatible with the 'polluter pays' principle, but fails to account for the differences in historical responsibility for the current stock of atmospheric CO_2. One possible solution is to reallocate the tax revenues and compensate those adversely affected by the implication of the taxation. More generally, there may also be concerns over national sovereignty and the influence a harmonised tax system may have on a country's ability to make its own domestic policy decisions.

6.3.2 Global cap and trade system

In a global cap and trade system, an overall restriction is set on the quantity of emissions allowed over a given period. Within this overall limit, emitters are allocated quotas of emissions rights that they can trade with each other as a commodity. The defining characteristic of a cap and trade market is that scarcity, and thus the value of carbon permits, is created through government intervention. National cap and trade markets can be created through domestic regulation, whereas international cap and trade markets require negotiation between participants.

Cap and trade is potentially highly effective at reducing emissions.[15] It provides certainty over the global emissions that are reduced which can be directly linked to the science underpinning the emissions reductions target (see Box 6.2). This is in contrast to an inflexible taxation system, where the cost is certain, but due to imperfect information the emissions reductions are not.

As long as coverage is global and all countries participate, cap and trade can address the criteria of leakage, additionality and permanence. International coverage means that all emissions should be accounted for. Therefore inherently there will be no leakage. Additionality depends on the level of ambition when setting the cap and the national inventory that it will be measured against. As with taxation, permanence within a cap and trade system relies on the durability of the system as well as the continued participation of all countries. However, the financial incentives provided to countries by a global cap in the form of liabilities and rewards generated encourage long-term participation.

The global cap and trade system performs well against the efficiency criterion because the combination of universal coverage and the ability to trade provides flexibility over the location of emissions reductions. As with taxation, all of the market participants are faced with a uniform carbon price, but the price in this system is set indirectly by the stringency of the cap and the cost of abatement. The total scale of funding automatically adjusts to the level required to produce sufficient abatement to meet the cap. This allows abatement to occur where it can be done most cheaply.

---

14 Mackenzie (1990)
15 Stern (2008)

> **Box 6.2: The economics of a cap and trade system**
>
> Under a cap and trade mechanism, an overall cap on emissions is established for a country or region. A global cap can be based on a global emissions stabilisation target. The total emissions below the cap are then allocated or auctioned among emitters (eg power plants, industrial plants, transport etc), who are then free to trade their allowances. Trading provides the most efficient means of meeting global reductions at lower cost to the international community.
>
> Figure A below illustrates how the price and quantity of emissions allowances is set in a cap and trade system.
>
> **Figure A:** Emissions price and quantity in a cap and trade system
>
>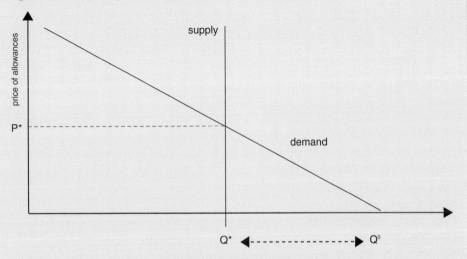
>
> The supply curve is vertical because once the allowances have been distributed no more will be created regardless of the price reached.[16] The demand curve is downward sloping because a rising carbon price will make it more attractive to reduce emissions rather than buy allowances. The price (P*) and quantity (Q*) of allowances is set by the point where the supply and demand curves intersect. Business as usual emissions, assuming no carbon price, are represented by the point Q^0. Therefore, the total quantity of abatement, or emissions reductions, is determined by the distance between Q* and Q^0.

A possible exception to this is if the scheme includes price-floors, price-ceilings, or enforcement mechanisms that have a similar effect (eg fines for non-compliance). In this case, it is no longer guaranteed to reach a given scale of funding, but the total costs are bounded. If prices prove volatile, and the costs to businesses of such volatility are high, then price caps may increase efficiency. However, firms habitually deal with volatile core input prices, or exchange rate changes, and will develop futures to hedge against predict-

---

16 There is, however, the issue of credibility; ex ante promises to limit the supply of allowances may not be kept if the permit price rises too high ex post.

able risk. Moreover, as authorities gain experience with cap and trade systems, they can be expected to smooth volatility, much as they do in the management of bond markets. If abatement costs are known, it is possible to achieve the same result with taxation and fully auctioned trading. However, given imperfect information, it is highly unlikely that the costs would be known to any reasonable level of accuracy ex ante (see Chapter 5 on mitigation costs).

A major advantage of global cap and trade over taxation lies in its distributional flexibility. This means that the system can be designed to be more equitable, particularly when considering developing countries. It is possible to achieve almost any distributional outcome when dividing up the allocation of allowances. This flexibility maximises the chance of finding a politically acceptable and equitable distribution of the burden of emissions reductions. For example, developing countries with little historical responsibility for climate change can receive very loose caps (based on need) while developed countries could take on the bulk of the target. Furthermore, a system to address the practical difficulty in agreeing differentiated national emissions targets required for a cap and trade system has already been established through the Kyoto negotiations.[17] The option of devolved, or company-based, trading schemes on a national or regional level enhances the equitability of a cap and trade scheme because the emitter, such as a power station, is directly liable for CO_2 released into the atmosphere.

In summary, a global cap and trade system performs best against the criteria of effectiveness, efficiency and equity for the following reasons:

- It places an absolute limit on total emissions and provides a direct link with the science underpinning the emissions target.
- It encompasses all countries and sectors and facilitates mitigation where it can occur most cheaply.
- Countries would be free to meet their cap through domestically determined policies, ie through a mixture of taxation, regulation and the purchase of international credits.
- It could encourage international consensus through the flexibility in dividing the cap in an equitable way.

An internationally harmonised taxation system could be an efficient tool as theoretically it equalises abatement between sectors and regions, but it would encounter severe difficulties in achieving effectiveness and equity:

- It does not provide a quantitative limit on the overall level of emissions required.
- It becomes administratively complex and costly to account for the common but differentiated responsibilities principle.
- Above all, there would be a practical difficulty in gaining the level of political commitment and international consensus that would be required to agree minimum standards across all sectors and countries.

This does not mean that taxation does not, or should not, play an important role in reducing carbon emissions. Domestic taxation and cap and trade can operate in parallel. Furthermore, as later chapters of the Review demonstrate, carbon markets are likely to take time to grow, leaving a funding gap in the short to medium term, particularly for a

---

17   Frankel (2007)

sector such as forests which is a relatively large source of emissions in developing countries. Nonetheless, a cap and trade system is an essential tool if the international community is to meet its climate change targets.

6.4 Rationale for including forests within a global cap and trade system

Integrating forests within a global cap and trade system would create opportunities to tackle a large part of current CO_2 emissions while delivering substantial finance for forest conservation and sustainable forest management. Excluding the forest sector would impede the benefit of trading to maximise emissions reductions and minimise costs.

We have already established that a global cap and trade system has the potential to deliver the emissions reductions required for global climate stabilisation but that it will maximise both sectoral and overall efficiency only if the market includes all relevant sources of emissions, including forests. However, the specific challenges that forests present mean that concern has been raised about the full integration of forests within a cap and trade system. While accepting the need for a cap and trade system for other sectors, some proposals for funding forest emissions reductions have opted to treat forests separately. For this to occur, an alternative source of funding would need to be similar to a cap and trade system in terms of scale, effectiveness, efficiency and equity. Multilateral funding in the form of donations from industrialised countries has been proposed as an option.

In Chapter 5 the average annual cost of mitigation for halving deforestation to 2030 is estimated to be between $17–33 billion per year. The Organisation for Economic Co-operation and Development (OECD) estimates that the total overseas development assistance (ODA) and official aid (OA) to forestry by OECD countries and multilateral agencies was an annual average commitment of $564 million between 1996 and 2004.[18] Only part of this funding would be directed to reduce emissions from deforestation. Given the magnitude of funding required, therefore, a system that separated forests from the carbon market and instead financed a reduction in forest emissions solely through multilateral funds would be highly unlikely to reach the required level of funding.

Furthermore, even if the level of funding were to be scaled up, the additional demand for finance for forests might risk a depletion in overall funds available to address other aspects of climate change in developing countries such as adaptation and technology. This does not mean, however, that multilateral funds cannot play an important part in an overall international response to forests. Chapter 13 will discuss possible public funding options for forests, including specific forest-dedicated funds, as a supplementary mechanism to the cap and trade system in the short and medium term.

Ultimately, excluding emissions from the forest sector would frustrate the ability of trading to maximise global emissions reductions and minimise the cost of achieving our overall climate goals. There would be an unexploited opportunity to abate an extra tonne of emissions in one sector compared with another. This is because there would no longer be a uniform price signal, and consequently there is the risk of having too much or too little abatement from other sectors.

---

18 Tomaselli (2006)

6.4.1 Modelling the effects of including forests in a cap and trade system

Including reduced emissions from deforestation and degradation (REDD) in a well-designed cap and trade system could reduce deforestation emissions by up to 75 per cent in 2030. With the addition of afforestation, reforestation and restoration, this would make the forest sector carbon neutral.

Including forests in a global cap and trade system would mean that the cost of halving global carbon emissions from 1990 levels could be reduced by up to 50 per cent in 2030 and up to 40 per cent in 2050.

This could allow the international community to meet a more ambitious global stabilisation target. By 2050, CO_2 emissions could be reduced by an additional 10 per cent with the inclusion of the forest sector.

Forest carbon finance could also make a significant impact on reducing poverty through increased financial flows to developing countries.

New modelling was commissioned for this Review from the UK Office of Climate Change Global Carbon Finance (GLOCAF) model to look at the effects of including forests on overall global emissions reductions and costs in a global cap and trade system.[19] The volume of abatement from the forest sector generated by a cap and trade market depends on the stringency of the cap and the cost effectiveness of the forest sector relative to other sources of abatement. If there is a loose cap or other abatement opportunities are cheaper, a cap and trade system will not generate a significant amount of finance for reducing forest emissions.

Several possible scenarios were examined using the model. The reference scenario assumed that the world adopts a global cap and trade regime covering all sectors apart from the forest sector by 2030. The cap corresponds with a 475ppm stabilisation scenario. This level of effort is consistent with a 50 per cent reduction in global emissions by 2050 compared with 1990 emissions (see Figure 6.1). This is based on an interpretation of an agreement by G8 leaders at Heiligendamm. Annex I countries were assumed to take on the majority of reductions; and all countries were assumed to take on binding targets of varying stringencies by 2030, in order to meet the global target. The reference scenario excluded forests from the cap and trade system, with forest emissions expected to continue on a business as usual path, assuming no other policies to address deforestation. The reference scenario result was then compared with the impact of including the forest sector in the scenario of a global cap and trade system in 2030 and also 2050. The model takes into account REDD[20] as well as ARR activities.

Under the reference scenario, emissions from deforestation were estimated to be 3.5$GtCO_2$ per year by 2030. By including forests in the global cap and trade system, emissions from deforestation were projected to fall to 0.9$GtCO_2$ per year by 2030. This reduction of 2.6$GtCO_2$ per year represents a 75 per cent reduction in emissions from deforestation by 2030.[21]

---

19 See www.occ.gov.uk/activities/gcf.htm
20 The model is calibrated from IPCC Fourth Assessment (2007) figures which also include forest degradation. For an explanation of how degradation is defined see Chapter 2.
21 These estimates were generated through the DIMA model. See Chapter 5 for further explanation of the DIMA model.

Furthermore, if ARR is included alongside REDD in the global cap and trade system, an additional 0.9GtCO$_2$ is projected to be sequestered. Consequently, with REDD and ARR in a global cap and trade system by 2030, the forest sector as a whole could be carbon neutral; ie, the amount of carbon released into the atmosphere is balanced by the amount of carbon sequestered.

The modelling was also used to determine the effect of including forests in a cap and trade system on the costs of achieving significant emissions reductions. The model projected that the cost of halving global carbon emissions from 1990 levels could be reduced by between 25 per cent and 50 per cent in 2030 and between 20 per cent and 40 per cent in 2050 if the forest sector is included (see Figure 6.3).

Figure 6.3: Global costs from excluding forests

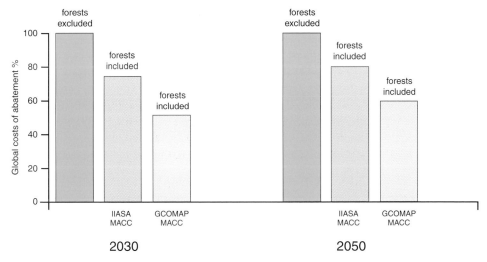

Source: GLOCAF Modelling for the Eliasch Review.

The significant effect of including forests on the costs of achieving substantial emissions reductions is related to improved efficiency. The exclusion of forests from a cap and trade system would mean the emissions target must be met by abatement from other sectors. Forests could supply potentially cheap carbon credits to the market. Therefore, their exclusion significantly increases the costs and inefficiency of achieving an overall emissions reduction target.

Under a reference scenario, with forests excluded from the system, total emissions in 2050 were projected to be reduced to 20GtCO$_2$ at a cost of 3 per cent of global GDP. For the same cost, including the forest sector (ARR and REDD) would potentially reduce CO$_2$ emissions to 18GtCO$_2$ in 2050, representing a 10 per cent reduction on the baseline scenario. This indicates that, by including the forest sector in a cap and trade system, the international community could set and meet a more ambitious global stabilisation target.

Forest carbon finance generated by the inclusion of forests in a global cap and trade system could also make a significant impact on reducing poverty if designed well. Countries with the highest potential to gain from reduced deforestation are predominantly developing countries, including many of the poorest countries in the world.[22] Therefore, cap and trade system, inclusive of forests, has the potential to increase financial flows from rich to poor countries.

The level of finance flowing to forest nations will depend on target setting, robust measuring and monitoring, efficient design of the carbon market and good governance. Some of these will be determined by international negotiations that will require incorporating all sectors and all countries. However, as an indication of the potential financial flows from rich to poor countries under a global cap and trade system, GLOCAF modelling for this Review suggests that these flows could be highly significant. For example, the modelling estimates that the forest sector has the potential to generate financial flows to sub-Saharan Africa of over $15 billion by 2030.

These significant financial flows are estimated because, with a global carbon market, countries with a lot of cheap abatement (typically developing countries) can reduce their emissions to meet their target and make a profit at the same time. For developed countries it is cheaper to buy abatement from abroad and give developing countries a profit than it is to do expensive abatement at home. In this way a carbon market can reduce the costs to developed countries and at the same time create profits for developing countries. It is important to note, however, that achieving these financial flows and ensuring that the finance actually results in reduced emissions while benefiting indigenous communities will require significant support and capacity building in many countries. Chapters 12 and 13 of this Review discuss governance and distribution of finance in more detail.

6.5 Four key elements of a long-term framework

In order to provide the international community with a guarantee of climate stabilisation within safe levels, there needs to be a clear and ambitious international target. Although target setting is a political decision, it will need to be based on the science and should be informed by an assessment of the likely costs of abatement options. Additionally, targets will require an equitable division of the global cap. The process of setting effective targets using the method of baselines will be described in detail in Chapter 9.

Once targets or baselines are set, robust measuring and monitoring techniques are needed to determine the reduction of emissions against a reference level. National inventories for all emissions sectors carried out on a consistent basis, covering all sectors and countries and with transparent verification, will be important within a cap and trade market. International inventory methods for the forest sector have been developed by the IPCC. Application of these methods is not only essential in estimating emissions, removals and carbon stocks but is also part of the system to guard against leakage, non-additionality and impermanence. This is discussed further in Chapter 10.

The design of any international carbon trading framework will need to generate net financial transfers from industrialised to developing countries. One of the concerns that has been expressed about forests being part of a single carbon market is that, because it is a potentially abundant source of cheap abatement, it would flood the market and lead to a collapse of the carbon price unless its integration was well designed. This is an impor-

---

22 Ebeling and Yasue (2008)

tant transitional issue and requires consideration of the relationship between supply and demand in the carbon market as well as appropriate design features at the point of linking forests to carbon markets. This is discussed in more detail in Chapter 11.

Carbon finance could make a significant impact on poverty reduction and loss of ecosystem services if governance is addressed within the scheme's design. Effective national institutions that are able to administer the system and enforce compliance are essential to ensure the confidence of the international community and the markets. Policies and international conventions that protect the forests and involve the poor are also key components. This is discussed in more detail in Chapter 12.

6.6 Conclusion

Various systems exist for achieving reductions in emissions from the forest sector as part of an overall global framework. A range of taxation options exists, both in terms of domestic taxation in forest nations and flows of international funds from developed countries raised through tax revenues or other means. Many of these are potentially valuable tools in tackling climate change. However, as a long-term framework, a cap and trade system that includes forest sector net financial flows from industrialised to developing countries performs best against the criteria of effectiveness, efficiency and equity.

Providing coverage is global and all major countries participate, cap and trade can address the challenges of leakage, additionality and permanence. International coverage means that all emissions should be accounted for with no leakage. Additionality depends on the accuracy of the cap-setting process and the national inventory that is used to measure and monitor real reductions in emissions. Permanence of emissions reductions relies on the durability of the system as well as the continued participation of all countries. The financial incentives provided to countries by a global cap in the form of liabilities and rewards generated encourage long-term participation.

The success of a cap and trade system, however, will depend on how well it is designed. Four key elements will be required for all sectors, including the forest sector: effective national targets; robust measuring and monitoring; an efficiently designed carbon market; and good governance. This chapter shows that, with these elements in place, the forest sector could become carbon neutral in 2030. Furthermore, the inclusion of the forest sector in an overall framework could increase the ambition of a global stabilisation target and/or reduce the global cost of meeting it.

While this chapter sets out a long-term framework that the international community could work towards, it is important to understand how the current international framework compares with this goal. The following chapter examines the current climate change system and reviews its successes and limitations.

7. The current international climate change framework

Key messages

Implementation of the current international climate change framework is a long way from delivering the emissions reductions required for a global stabilisation target.

The United Nations Rio Conventions established the importance of addressing forests on an international scale. The United Nations Framework Convention on Climate Change proposes sound principles for international action to tackle the challenge posed by climate change.

The Kyoto Protocol of the UNFCCC commits developed countries to legally-binding targets to limit or reduce their greenhouse gas emissions, and provides mechanisms that allow emissions trading.

Targets for the first commitment period of the Kyoto Protocol have been recognised as insufficient to meet the reduction in total anthropogenic emissions required for a global stabilisation target of 445-490ppm CO_2e. Further action will be needed from developed and developing countries to meet this goal, with responsibilities and rewards for reducing emissions, including those from the forest sector.

Although developed countries are required to estimate and report emissions from land-use, change and forestry annually, and developing countries do so periodically, forest emissions may not be comprehensively estimated by countries, and the forest sector has tended to lag behind other sectors in measuring and monitoring.

While Annex I countries are able to use the three Kyoto mechanisms to achieve their commitments, non-Annex I countries are only able to benefit from hosting afforestation and reforestation projects under the Clean Development Mechanism. Reduced deforestation and degradation is not included in the international system for developing countries.

Accreditation within the CDM, particularly for forest projects, is widely recognised as needing reform. Transaction costs are substantial and the temporary nature of A/R credits have had significant impacts on the uptake of A/R CDM projects.

7.1 Current international action

Implementation of the current international climate change framework is a long way from delivering the emissions reductions required for a global stabilisation target.

The preceding chapter identified the key advantages of a comprehensive global cap and trade system to address climate change and to deliver a reduction in emissions in an effective, efficient and equitable manner in the long term. If the benefits of a cap and trade system are to be fully realised by maximising global emissions reductions and minimising the cost of achieving our targets, it is important that forests, along with all other emissions sources, should be included.

This chapter considers the current institutional framework and the international agreements that have so far attempted to tackle the effects of climate change. It outlines the sound principles for international action on climate change provided under the UNFCCC, before assessing the implementation of the framework. The current international response to climate change has limitations: notably it does not include emissions commitments from all sectors and all major emitting countries, and there is a clear need for substantially deeper commitments in the future.

The chapter focuses on the real and perceived differences between forests and other sectors that have presented challenges not yet comprehensively addressed in climate change negotiations; and on the absence of meaningful incentives for developing countries to undertake mitigation of forest sector emissions. These issues need to be addressed so that non-Annex I countries (mostly developing countries) and forest countries can benefit from a future climate change agreement in similar ways to Annex I countries.

7.2 The United Nations Rio Conventions

The United Nations Rio Conventions established the importance of addressing forests on an international scale. The United Nations Framework Convention on Climate Change proposes sound principles for international action to tackle the challenge posed by climate change.

Achieving consensus and harnessing action on environmental issues at any level is not easy. The fact that broad international agreement has been reached over the action required to address climate change and sustainable development is correspondingly impressive. The Rio Conventions and associated agreements achieved international consensus and thus form the first leading multilateral environment agreements on global environmental protection (see Box 7.1).

Box 7.1: Forests and the United Nations Rio Conventions

The Conventions and associated agreements launched at the Rio Earth Summit in 1992 put in place an international framework to tackle a wide but inter-linked set of environmental problems. The Conventions have a particular relevance to forests because forests provide benefits and services that are key to all the Conventions: safeguarding biodiversity, protecting ecosystems, sequestering carbon and preventing its release into the atmosphere.

The UN Convention on Biological Diversity (CBD) was the first global agreement to combine all aspects of biodiversity.[1] Countries are required to develop national strategies for the conservation and sustainable use of biological diversity, including the creation of a system of protected areas to conserve that diversity.[2] Forests are addressed directly in the CBD through the programme of work on forest biodiversity, which includes biophysical aspects, institutional and socio-economic environments and forest monitoring.[3] CBD negotiations on access and benefit sharing (ABS), which are due to conclude in 2010, will also have significant implications for the management of forest resources.

The UN Convention to Combat Desertification (UNCCD), which was agreed in 1994 in pursuit of a mandate from Rio, obliges those countries affected by desertification to implement national and regional action programmes to tackle desertification and mitigate the effects of drought. Developed countries are expected to mobilize substantial new and additional funding to help developing countries fulfil their obligations, in addition to providing access to appropriate technology and know-how. Key objectives of the UNCCD are the prevention and reduction of land degradation, the reclamation of desertified land, and the rehabilitation of partly degraded land, in which forestation (afforestation, reforestation and forest restoration) is a key tool.[4]

The UNFCCC entered into force in 1994 and establishes a robust and comprehensive framework for intergovernmental efforts to tackle the challenge posed by climate change. Its credibility is demonstrated through the participation of nearly all countries in its decision-making. The Convention makes clear that developed countries should take the lead in combating climate change and that all parties to the Convention need to take precautionary measures to mitigate the adverse effects of climate change.[5]

The UNFCCC sets specific aims for developed countries to return their greenhouse gas emissions to 1990 levels by 2000, with more general commitments established for all parties. Although mandatory emission caps for individual nations were not specified by the UNFCCC itself, parties are committed to produce, and regularly update, national inventories, setting out the amount of greenhouse gas emissions released and how much is absorbed by sinks. The Kyoto Protocol to the UNFCCC, which was agreed in 1997 and entered into force in 2005, sets mandatory emissions reduction or limitation commitments for developed countries for the first commitment period, 2008 to 2012,

---

1   United Nations (2002)
2   United Nations (1993). (CBD)
3   COP 6 Decision VI/22 (April 2002)
4   United Nations (1994). (UNCCD)
5   United Nations (1992). (UNFCCC)

> and sets up a process for negotiating further commitments for subsequent commitment periods.[6]
>
> In addition to the three conventions, the 1992 Rio Summit agreed the Rio Declaration on Environment and Development (known as the Rio Principles) and Agenda 21, a comprehensive programme for delivery. These have established the basis for subsequent international consideration of sustainable development in general. The Forest Principles agreed in Rio[7] set out non-legally binding principles for the management, conservation and sustainable development of forests. They led to the establishment of the United Nations Forum on Forests (UNFF) which in 2007 agreed to strengthen the political commitment of UN member states to the sustainable management of all types of forests.

While achieving international consensus, the Conventions also illustrate that further elements will be needed to deliver significant emissions reductions in the future. For example, they have faced criticism for inadequate resourcing and a lack of real commitment from parties to follow through the promises they made.[8] The Conventions, as individual entities, present challenges in their coordination, especially as responsibility for each Convention does not always lie with the same institution at the national level.[9] Countries have recognised this as an issue and have recently established an Ad Hoc Technical Expert Group (AHTEG) to facilitate coordination between the UNFCCC and the CBD.

The UNFCCC, however, takes important steps towards the implementation of key elements to address significant emissions reductions. It establishes an 'ultimate objective' for all parties: the '...stabilisation of greenhouse gas concentrations in the atmosphere at a level that would prevent dangerous anthropogenic interference with the climate system.'[10] Particularly important is the requirement to produce and submit to the Conference of Parties (COP – see Box 7.3) a national inventory of greenhouse gas emissions by sources and removals by sinks.

While acknowledging that tackling climate change requires global commitment and action, the convention separates countries into distinct categories:

- Annex I (OECD countries and economies in transition);
- Annex II (OECD countries only);
- non-Annex I (mostly developing countries).

This establishes the principle of 'common but differentiated responsibilities', meaning that the effort required to tackle the effects of climate change should be equitably divided.[11] Developed countries should accordingly take the lead, as they bear the greatest historical responsibility for the current levels of GHG emissions in the atmosphere, and have greater capacity to address them. This is reflected in the more detailed commitments for developed countries and an obligation to provide 'new and additional financial resources' to developing country parties, which are the most vulnerable to climate change.

---

6 United Nations (1998)
7 United Nations (1992). (UNCED Report)
8 World Wildlife Fund (2002)
9 Hoffman (2003)
10 United Nations (1992). (UNFCCC)
11 Yamin and Depledge (2004)

Table 7.1: Components of the UNFCCC and Kyoto Protocol

Key component	Country parties, obligations	Combined implication for forests
Targets		
UNFCCC	Annex I countries committed to taking measures to limit anthropogenic emissions and enhance sinks with the aim of returning emissions to 1990 levels.	Annex I: Afforestation, reforestation minus deforestation must be accounted for. Other land use, land-use change and forestry (LULUCF)[12] activities such as forest management can be elected. CDM forestry projects can be used to meet 1% of base year emissions for Annex I countries.
Kyoto	Individual legally-binding caps for Annex I countries based on reduction or limitation of emissions relative to the level in a base year, generally 1990. Emissions targets must be met by the commitment period 2008-2012.	
Inventories		
UNFCCC	Commitment to develop, update and publish national inventories for all Parties, and annually for developed countries.	Annual reporting of emissions and removals by Annex I countries, including emissions and removals from LULUCF, whether or not used to meet commitments. Developing countries provide inventory data periodically, with national communications.
Kyoto	Obligation for Annex I countries to submit an inventory each year. Non-Annex I countries to submit an inventory periodically. IPCC methodology should be used to measure emissions.	

---

12 LULUCF is now referred to as Agriculture, Forestry and Other Land Use (AFOLU) within the IPCC Guidelines for National Greenhouse Gas Inventories (2006).

Key component	Country parties, obligations	Combined implication for forests
Funding		
UNFCCC	Developed countries provide financial resources to developing countries to meet the costs of compiling national inventories, communications and new technology. The Global Environment Facility serves as the financial mechanism (see Box 7.3)	Afforestation and reforestation projects (A/R) can be funded through CDM by Annex I countries. No provision for funding of reduced emission from deforestation and degradation (REDD) projects under CDM.
Kyoto	Reinforcement of developed countries commitment to meet the full costs incurred by developing countries of implementing their commitments. CDM projects funded by Annex I countries, hosted in non-Annex I countries	
Emissions trading		
UNFCCC	No detailed provisions for emissions trading	Removal units (RMUs) from sequestration activities can be traded. CDM allows A/R, but not REDD projects in non-Annex I countries to be credited and traded as expiring credits.
Kyoto	Mechanisms to enable trading established.	
Sustainable development		
UNFCCC	All countries committed to promote sustainable management and conservation of ecosystems	For Annex I countries afforestation, reforestation and deforestation (ARD) activities incentivised but management of pre-1990 forest can only be used to a limited extent with country specific caps. CDM forestry projects should promote sustainable development
Kyoto	Objective of CDM: to help non-Annex I Parties achieve sustainable development while observing their national sovereignty. The DNA* of a non-Annex I country assesses the effects of potential projects before approval can be granted.	

* DNA – Designated National Authority is the body granted responsibility by a party to authorise and approve participation in CDM projects. Approval must confirm that the project activity contributes to sustainable development in the country

7.3 The importance of the Kyoto Protocol

The Kyoto Protocol of the UNFCCC commits developed countries to legally-binding targets to limit or reduce their greenhouse gas emissions, and provides mechanisms that allow emissions trading.

7.3.1 Commitments under the Protocol

While the UNFCCC sets the framework for international action to address climate change, the 1997 Kyoto Protocol to the UNFCCC significantly strengthens the objective, principles and institutions of the convention in the areas of binding targets, measuring and monitoring and the introduction of international trading (see Table 7.1).

Binding caps

The major development represented by the Protocol is that, whereas the Convention encouraged industrialised countries to stabilise greenhouse gas emissions, the Protocol commits them to reduce or limit emissions.

The Kyoto Protocol commits Annex I countries (39 developed countries and countries with economies in transition) to individual, legally-binding emissions reduction targets, called Quantified Emission Limitations or Reduction Commitments. Collectively these targets were intended to amount to a minimum of 5 per cent against 1990 levels over the first commitment period 2008-2012, with some nations committing to higher percentage reductions and some lower. Despite the relatively modest scale of emissions reductions, the fact that any form of binding national emissions restraints was agreed represents significant progress towards more ambitious target setting.

In meeting these commitments, land use, land-use change and forestry (LULUCF) are treated somewhat differently from other sectors. Under the accounting provisions of Article 3.3 of the Kyoto Protocol afforestation, reforestation and deforestation (ARD) activities since 1990 must be accounted for. However, under the provisions of Article 3.4, carbon stock changes that relate to the management of forests in existence prior to 1990, as well as carbon stock changes from cropland management, grazing-land management and revegetation, only need to be included for the first commitment period if countries choose to do so.

National inventories

The Protocol represents a significant development in the comprehensive coverage of measuring and monitoring. All Annex I countries must submit information on greenhouse gas emissions and removals by sinks on an annual basis, whatever choices they have made under the provisions of Article 3.4 (see Table 7.1). This information must include activities in the LULUCF sector. The information is prepared using internationally agreed guidance from the IPCC, which aims to cover all significant categories of anthropogenic emissions, including the LULUCF sector.

The information submitted by Annex I countries is subject to annual review by an expert international review team coordinated by the UNFCCC Secretariat, which reports back to the parties. If a country emits more than its allowance – taking account of any international transfers under the flexible mechanisms – it has to make up the difference in the next commitment period, plus an additional deduction of 30 per cent. Kyoto Protocol obligations on non-Annex I countries include producing, publishing and updating national – and where appropriate regional – programmes containing measures both to mitigate climate change and adapt to it, but without any quantified emissions reduction or limitation commitments.

Emissions trading and the Kyoto mechanisms
Although Annex I countries should meet a significant part of their legally-binding cap through domestic effort, the Kyoto Protocol includes the provision for countries' obligations under the international framework to be met flexibly according to their national circumstances, respecting national sovereignty. For example, parties are able to pay other countries or entities to reduce emissions, or offset emissions abroad, where the costs could be lower. Under Kyoto, Annex I countries are allocated tradable emissions rights in the form of 1 tonne units of CO_2e up to the quantity of emissions allowed by their commitment (this allocation is called the assigned amount). They are then able to transfer or acquire these units to or from other Annex I countries through the International Transaction Log (see Box 7.3). Thus units acquired from one party are added to the assigned amount for the acquiring party. This establishes the basis of emissions trading.

The introduction of the emissions trading system has provided incentives for actions to reduce or mitigate emissions through a variety of methods including regional initiatives, unilateral state initiatives, bilateral initiatives, cooperation between sub-national governments, and international private sector and public-private partnerships.

Although meeting commitments remains a national responsibility countries may, in effect, pass on (or devolve) emissions allocations to companies directly responsible for the emissions, such as power plants. In order to stay within their cap, the company has to either reduce its emissions or pay other companies or offset providers for additional units. The most important current example of translating sovereign obligations into domestic policies is the EU Emissions Trading Scheme (EU ETS). This and other regional carbon markets are discussed further in Chapter 11.

The Kyoto Protocol establishes three mechanisms: International Emissions Trading, Joint Implementation and the Clean Development Mechanism, which collectively constitute a prototype international emissions trading framework or carbon market (see Figure 7.1 and Box 7.2).

Under International Emissions Trading (IET), Annex I countries that are over their emission targets are permitted to buy allowances or credits from other Annex I countries that have emissions to spare (see Box 7.2). Under the Joint Implementation (JI) mechanism, an Annex I country (Country A) with an emissions reduction commitment may (with agreement) implement an emission-reducing project or a project that enhances removals by sinks in another Annex I country (Country B), also with an emissions reduction commitment. Country A can then count the resulting emission reduction units (ERUs) from Country B towards its own Kyoto target.

Under the CDM, Annex I countries are, with the agreement of the developing country, able to implement project activities that reduce emissions in developing countries (non-Annex I parties), in return for certified emission reductions (CERs) or in the case of forestry projects, temporary or long-term CERs (tCERs or lCERs respectively). Temporary credits are a means of dealing with the permanence risk of project-based activities in the forest sector.[13] These project activities must assist non-Annex I parties in achieving sustainable development objectives.

Figure 7.1: The three Kyoto mechanisms

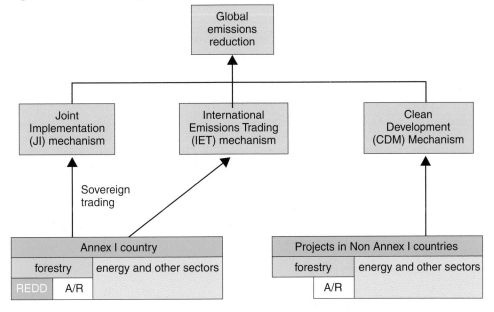

Box 7.2: Credits traded under the three Kyoto mechanisms

The reduction targets set for each developed country are expressed in terms of a commitment. Participating countries are allocated emissions rights up to the quantity of emissions allowed by their commitment. These allowances can then be traded with other Annex I countries. Additionally, Annex I countries can obtain certified credits from emissions reductions via Joint Implementation or the CDM.

International Emissions Trading Mechanism (IET)
Annex I countries that exceed their emission targets are permitted to buy credits from other Annex I countries that have excess emissions rights to meet their commitment. There are two types of credits that can be traded in this way. Assigned Amount Units (*AAUs*) are tradable sovereign allowances to emit CO_2e. AAUs are issued by parties to the Kyoto Protocol to Annex I countries, calculated by reference to their base year emissions and their emission reduction commitment, or cap. Removal Units (*RMUs*) are tradable sovereign sequestration credits generated from land use, land-use change and forestry (LULUCF), ie reforestation and afforestation activities, calculated against national baselines.

---

13 Section 7.4 discusses temporary, or expiring credits in more detail.

Joint implementation Mechanism (JI)
JI is a project-based mechanism for Annex I countries. Under JI, an Annex I country may invest in an emission-reducing project or a project that enhances removals by sinks in another Annex I country. The credits earned from a JI project can be counted towards meeting the Kyoto target of the investing country. The private sector can be authorised to take part in JI projects. Credits generated by JI projects are Emissions Reduction Units (*ERUs*). ERUs are created by deducting RMUs or AAUs from the host country's registry account so that they can be credited to the investor's registry account.

Clean Development Mechanism (CDM)
CDM is a project-based mechanism to assist non-Annex I countries to achieve sustainable development and to help Annex I countries achieve their emissions reduction and limitation commitments. To help meet its Kyoto target, an Annex I country may implement project activities that reduce emissions in non-Annex I countries. The tradable credits generated by non-forestry CDM are called Certified Emissions Reductions (*CERs*)

CDM crediting can also take place for forestry activities. However, this only applies to afforestation and reforestation activities. Reduced emissions from deforestation and degradation (REDD) activities are excluded from crediting.

Afforestation and reforestation projects under the CDM generate temporary and long-term CERs (*tCERs* and *lCERs* respectively) that are equivalent to one metric tonne of CO_2e but only for a finite period. A tCER expires at the end of the commitment period following the one in which it is issued, while an lCER expires at the end of the crediting period of the afforestation or reforestation project activity under the CDM for which it was issued.

Box 7.3: Major institutional bodies under the UNFCCC

The United Nations hosts the key institutional bodies of the existing international climate change framework.

Conference of the Parties (COP)
The core institution of the UNFCCC is the COP, which every year brings together heads of state or their ministers from each of the 192 countries that have ratified the Convention. The COP provides, among other things, the institutional mechanism for executive decisions about the individual commitments that countries should undertake to help fulfil their common but differentiated responsibilities under the Convention. The Conference of Parties Serving as the Meeting of the Parties (the COP–MOP) serves similar functions for the Kyoto Protocol, which is a Protocol of the UNFCCC.

UNFCCC Secretariat
The UNFCCC Secretariat provides the institutional mechanism for technical and administrative work to support decision making within the COP and the numerous bodies and ad hoc working groups mandated to fulfil or elaborate its decisions. Under the UNFCCC Secretariat, the Subsidiary Body for Implementation (SBI) is charged with monitoring progress and performance, while the Subsidiary Body for Scientific and Technical Assistance (SBSTA) elaborates the scientific, technical and methodological rules for reporting emissions and mitigation measures in consultation with the IPCC.

> **Global Environment Facility (GEF)**
> The GEF is the financial mechanism of the UNFCCC and the other Rio Conventions. The UN plays a major role in executive control of the Facility. Programmes funded by the GEF are managed on behalf of the GEF Board by the World Bank.
>
> **International Transaction Log Administration (ITLA)**
> The ITLA is responsible for the administration of the International Transaction Log (ITL), an electronic platform for international trading via the Kyoto mechanisms. The administrator of the ITL itself is the UNFCCC secretariat. Whenever an emissions allowance or credit is issued, it is assigned a code and tracked by the ITL over its lifecycle, from issuance to expiry, cancellation or retirement.
>
> **CDM Executive Board**
> The CDM Executive Board is charged with putting the Kyoto commitments and COP decisions on the role and regulation of the CDM into practice. This includes adopting its operational rules, approving or rejecting proposed CDM projects and programmes, regulating the private companies authorised to verify the methodological rigour of project proposals, and auditing reported results.
>
> **Joint Implementation Supervisory Council (JISC)**
> The Joint Implementation Supervisory Council regulates and manages the approval process for Joint Implementation projects and programmes in countries that have not yet implemented certain national institutional functions. Those countries that have implemented these functions are exempt from JISC oversight, with regulatory responsibility performed by their national institutions.

7.4 Limitations of the first Kyoto commitment period

A full assessment of the Kyoto system must wait until 2014. The first commitment period of the Kyoto Protocol still has over four years to run, followed by a two year window for Annex I states to reconcile their accounts and trade against any remaining sovereign emissions liabilities. Nonetheless, it is instructive to consider the limitations of the Kyoto Protocol in order to guide our assessment of those components that need to be improved or replaced in order to support a more ambitious subsequent commitment period beyond 2012.

7.4.1 Emission targets

Targets for the first commitment period of the Kyoto Protocol have been recognised as insufficient to meet the reduction in total anthropogenic emissions required for a global stabilisation target of 445-490ppm CO_2e. Further action will be needed from developed and developing countries to meet this goal, with responsibilities and rewards for reducing emissions, including those from the forest sector.

Current emissions reduction commitments are a first step and are clearly insufficient to achieve climate stabilisation. The Kyoto Protocol therefore provides a process to negotiate future commitments. This is currently underway and due to complete in Copenhagen in 2009. Parties recognise the need for substantially deeper commitments; for example the EU has committed unilaterally to a 20 per cent reduction in emissions by 2020, and a 30 per cent reduction if a global agreement is achieved.

The emissions targets do not comprehensively cover all major emitters. Although a party to the UNFCCC, the United States failed to ratify the protocol following the expression of domestic concerns about its impact on national competitiveness.[14] And while the treaty was ratified by 181 countries, developing countries are not bound by individual targets. Only industrialised countries are required to limit their emissions, and only a subset of the most advanced industrialised countries (Annex I countries) are required to contribute towards the costs of building capacity in the less advanced countries.

Currently, developing countries account for about 50 per cent of energy-related carbon emissions, and their share is expected to rise to 70 per cent by 2030 in the absence of additional policies.[15] China, for example, currently emits about five tonnes of CO_2e per person, and India is approaching two tonnes of CO_2e. By 2050, eight billion out of a global population of nine billion people will live in developing countries.[16] Without commitments by a majority of countries, particularly the major emitters, to take on national targets in the long term, meeting the climate stabilisation target of 445-490ppm will not be possible.

7.4.2 Measuring and monitoring

Although developed countries are required to estimate and report emissions from land-use change and forestry annually, and developing countries do so periodically, forest emissions may not be comprehensively estimated by countries, and the forest sector has tended to lag behind other sectors in measuring and monitoring.

Commitments for measuring and monitoring of emissions also differ between Annex I and non-Annex I countries. For Annex I countries, inventories are subject to annual review. For example, Removal Units (for sequestration activities) cannot be issued unless national inventories have been checked and verified by expert review teams. National monitoring and verification of carbon sequestration is more effective than measuring and monitoring against a project baseline. It allows sequestration to be achieved through a number of different policies or activities as appropriate to the country in question. Effort can be delivered either through project level activities (tree planting, forest management techniques) or through indirect policy instruments (fiscal regime, construction standards, import-export regimes), which increases the scope for non-intrusive and cost-effective measures to generate emissions reductions.

The comprehensive approach to measuring and monitoring of emissions for Annex I countries (ie, all agreed emissions are covered) means that any risk of the impermanence of carbon is managed. Because of national estimation and accounting systems, deforestation and REDD would be no different to industrial emissions in this respect. In the event

---

14 The Bryd-Hagel Resolution of the US Senate (1997) which resolves that the US should not be a signatory states that 'the level of required emission reductions, could result in serious harm to the United States economy, including...trade disadvantages'
15 Stern (2008)
16 Stern (2008)

of a reversal subsequent to issuance of the credit, an emissions liability will be created in the seller's inventory.

The national measuring and monitoring system also addresses the concern for leakage (the displacement rather than absolute reduction of emissions) within a country. If emissions are measured at a national level, leakage from one place to another within a country does not affect the national emissions rate, so a country will not be credited.

Although national inventories are encouraged for non-Annex I countries and developed countries are committed to provide financial support to meet the costs incurred by non-Annex I countries, there is no obligation for non-Annex I countries to provide annual national inventories. As reductions in emissions from deforestation are not a credible asset under the CDM, the existing framework provides little incentive for non-Annex I countries to measure and monitor deforestation and REDD activities robustly at a national level, although countries may do this for national policy reasons.

Project-based measuring and monitoring of reforestation and afforestation activities is arguably inefficient. Crediting of CDM projects takes place against a project and not a national baseline, limiting the flexibility that is offered to Annex I countries domestically to achieve emissions reductions. Without a national commitment, displacement or leakage of emissions beyond project boundaries is particularly difficult to measure. This is largely the reason for REDD being excluded from the CDM in the first commitment period. The obligation to monitor carbon stocks lasts only as long as the life time of the project and therefore carries the risk of impermanence.[17]

7.4.3 Trading under the flexible mechanisms

While Annex I countries are able to use the three Kyoto mechanisms to achieve their commitments, non-Annex I countries are only able to benefit from hosting afforestation and reforestation projects under the Clean Development Mechanism. Reduced deforestation and degradation is not included in the international system for developing countries.

Annex I countries have access to all Kyoto mechanisms through which they can achieve their target reductions. However, because non-Annex I countries have not committed to a cap on their emissions, they are only able to host CDM projects. This means there are two separate climate change regimes, one for Annex I countries and one for non-Annex I countries (see Figure 7.2a and b).

International emissions trading mechanism
In terms of efficiency, it is difficult to assess the core international emissions trading mechanism (the mechanism for direct country-to-country trading of sovereign commitments) because there are relatively few trades prior to the reconcilliation of liabilities at the end of the first commitment period.

Because non-Annex I countries are not subject to sovereign trading, they are in a subordinate position in terms of access to the future global carbon market and devolved or company based emissions trading. Trade in carbon commodities within the devolved carbon markets is widely expected to increase markedly over coming decades. The value

---

17 Schneider (2007)

of the carbon market in 2007 ($64 billion) was over double that in 2006.[18] And it is projected to grow to $100 billion by 2030.[19] Developing countries have been kept on the margins of this potentially highly lucrative market.

Joint Implementation Mechanism

As with the international emissions trading mechanism, assessment of the JI also presents challenges. Credits under JI have not yet been issued and participation has been restricted to a small fraction of the countries and activities in which cost effective abatement opportunities are available. But there are strong indications of interest from the private sector to develop projects or programmes through JI. The pilot phase of the JI has also shown that, with the correct framework in place, investment could be substantial.[20] The JI mechanism also recognizes a wide variety of LULUCF projects activities such as avoided deforestation, forest and wetland management and sustainable agriculture as eligible for generating credits.[21]

Clean Development Mechanism

The sole mechanism that developing countries are able to benefit from is the CDM. The market aspect of the CDM can be seen so far as a clear success. As of April 2008, 1033 CDM projects have been registered and a total of 137 million CERs issued. Some 1250 million CERs are expected to be issued during the first commitment period.[22] Between 2003 and 2005 the CER quadrupled in price. However, it does not have the same comprehensive approach to forests as the JI and has failed to support any significant number of projects in the forestation sector, despite the large potential for cost effective mitigation in this sector in many developing countries.[23]

Furthermore, in terms of the geographical distribution of projects, and therefore the distribution of the financial benefit derived from the projects, the CDM has proved less successful. While the mechanism has successfully generated financial transfers from developed countries to developing countries the vast bulk of this finance has gone to middle income countries. China, for example, supplied 73 per cent of the market share in terms of 2007 transacted volume. The less advanced countries have supplied only a small percentage of the share of CDM credits. Africa as a whole supplied just 5 per cent of the market share.[24] Given their less advanced state of development, proposals in the least developed countries have tended to be for smaller projects of less interest to profit-oriented foreign investors seeking CDM investment opportunities. These countries are therefore at a particular disadvantage in mobilising finance required to overcome the CDM investment hurdle.[25] As a result, the countries where capital is genuinely scarce and the government most lacking in capacity have derived comparatively less benefit from the current CDM than the more advanced developing countries.

---

18 World Bank (2008)
19 UNFCCC (2007)
20 Michaelowa (2005)
21 Schlamadinger et al (2006)
22 Schlamadinger et al (2008) background paper
23 Schlamadinger et al (2008) background paper
24 World Bank (2008)
25 Michaelowa (2005)

Transaction costs for all CDM projects are generally high and they have been estimated to be up to 53 per cent of the total project costs.[26] The CDM regulatory process currently takes about 300 days, on average, from validation to registration.[27] Costs for approval of forestry projects can be excessive due to additional delays involved. The simplified modalities for small-scale A/R projects that had been developed to allow forest communities to participate in the CDM are still largely out of reach given the high institutional transaction costs of project preparation.[28]

A report commissioned by this Review examined the institutional and system barriers to undertaking forestry and other land use activities in non-Annex I countries, primarily through the CDM. The report highlighted the significant upfront costs required for reforestation projects and the delay before the first substantial credit units can be generated – usually at least five years as trees become established. The requirement for verification at five-yearly intervals can delay the certification of emissions reductions and associated returns still further.[29]

7.4.4 Accreditation

Accreditation within the CDM, particularly for forest projects, is widely recognised as needing reform. Transaction costs are substantial and the temporary nature of A/R credits have had significant impacts on the uptake of A/R CDM projects.

The Kyoto Protocol allows countries that have national emissions inventories to generate sequestration (from afforestation and reforestation) credits or RMUs against national sink sector data. The certification and registry processes for RMUs do not involve significant transaction costs because credits are certified and issued in large batches. RMUs also have high integrity because they are generated nationally and the relevant inventory data are audited by the UNFCCC, are under regular scrutiny and have common carbon accounting rules. This means that buyers can have a high degree of trust in the credits' robustness. If the RMU is traded with another country, the liability for its replacement in the event of a reversal subsequent to issuance, lies with the host government.

Forest credits generated by the CDM are very different in nature from RMUs. CERs are generated against project-specific baselines, and not national inventory data, where there is a presumption of long-term responsibility. In order to address concerns regarding potential non-permanence of CERs, the modalities and procedures for A/R allowed for the creation of one of two kinds of expiring CERs[30]: either a temporary CER (tCER) or a long-term CER (lCER). The tCERs expire at the end of the commitment period subsequent to the one in which they are issued, while lCERs expire at the end of the project crediting periods. Neither can be carried over to subsequent commitment periods (even if the carbon remains sequestered). Furthermore, responsibility for lCER replacement – if either a reversal or removals is detected at verification or if no certification is provided – falls to the party that has purchased the units.

---

26 Fichter et al (2003)
27 Stern (2008)
28 Robledo (2007)
29 Schlamadinger et al (2008)
30 Schlamadinger et al (2008)

While the creation of expiring units was considered a universally acceptable solution at the time, they have proved unattractive to investors. As part of the work commissioned for this Review, interviews were conducted with leading CDM A/R project developers. A common issue shared by those interviewed was the difficulty they encountered in securing investment for projects that are perceived to be of high risk and are not fungible to other CDM units. This has relegated A/R CDM credits to a substantially lower class of credit. It is estimated that tCERs are worth around 14-35 per cent of the value of other CERs, and lCERs potentially 45-100 per cent, depending on the discount rate and the crediting period length.[31] In practice both are estimated to be worth around 25 per cent of the value of standard CERs. Furthermore, the report suggests that there has been growing recognition that the risks inherent in A/R projects are not as great as previously thought and that alternative means of addressing non-permanence could be applied.[32]

The attractiveness of A/R credits has arguably been further limited through the decision to set a cap on their purchase. The Marrakesh Accords ruled that the use of eligible A/R project activities under the CDM by any one party could not exceed 1 percent of base year emissions. This decision was made because of concern that the large-scale availability of cheap forest credits would flood the market. This limit has been perceived as an additional barrier or risk to investment in an A/R activity.

Figure 7.2a: The current international climate change framework for Annex I countries

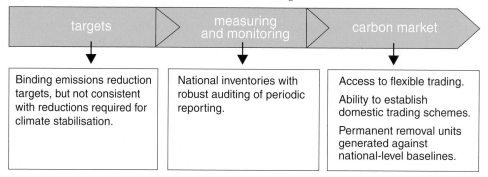

Figure 7.2b: The current international climate change framework for non-Annex I countries

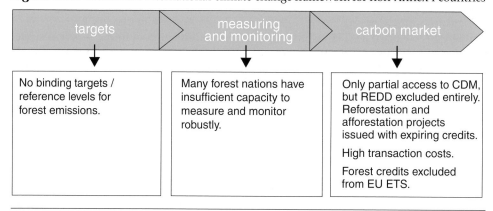

---

31 Dutschke et al (2005)
32 Schlamadinger et al (2008). Chapter 10 identifies alternative measures for addressing permanence while also providing the fungibility of credits.

7.5 Bali Action Plan

Emissions reduction targets for the first commitment period under the Kyoto Protocol were always recognised as a first step in delivering the reductions required for global climate stabilisation. The text of the Protocol recognises the need for subsequent commitment periods, and requires negotiations on future commitments.[33] This process was initiated at the 13th Conference of the Parties to the UNFCCC which took place in Bali in November 2007, and resulted in the adoption of the Bali Action Plan for formal negotiation of a new global deal. The plan recognises the need for urgent, meaningful action to reduce emissions from deforestation and forest degradation. The decision on deforestation in developing countries adopted at Bali encourages parties to 'identify options and undertake efforts, including demonstration activities, to address the drivers of deforestation.'[34] This is an important step towards the full integration of forests within the international climate change framework.

Underpinning the measures set out in the Bali Action Plan will be negotiation of the scale and distribution of international finance, in particular addressing 'enhanced action on the provision of financial resources and investment to support action on mitigation and adaptation and technology cooperation.'[35] This element is important for providing incentives for all countries, developed and developing, to participate in a comprehensive emissions reduction system.

Parties are currently working towards an agreement with deeper commitment and broader coverage, to be agreed at the 15th Conference of Parties in Copenhagen in 2009, with specific mention of the need to take action on deforestation and forest degradation, and the role of sustainable management of forests.

7.6 Conclusion

The Rio Conventions created an environment within which much learning has taken place. The Bali Action Plan, with its decision on reducing emissions from deforestation in developing countries, is a positive example of the commitment of many countries in the international community to exploring the role of forests in a long-term response to climate change.

Reform is needed to deliver the significant emissions reductions required. As this chapter identifies, the current agreements have several shortcomings. Targets are not sufficiently stringent; measuring and monitoring, particularly of forests, is not comprehensively implemented; and the trading mechanisms and the credits they generate have significant limitations for developing countries and the forest sector. The CDM entails high transaction costs, limiting access to the market of non-Annex I abatement potential. And current regulation of forest credits provides disincentives for wide participation.

---

33 United Nations (1998)
34 Decision 2/CP.13: Reducing emissions from deforestation in developing countries: approaches to stimulate action (2007)
35 UNFCCC (2007) (Decision 1/CP.13)

Chapter 6 presented a clear framework for a global cap and trade system that could address the challenges of delivering a climate change framework that is both efficient and equitable. The current international framework forms a sound basis for moving towards this long-term goal. In the meantime, urgent action is needed. A transitional system should be implemented in the short and medium term, to address the limitations of the current system and establish the four key elements that form the basis of the long-term framework. Chapter 8 considers the different options for a transitional system, and at how the limitations of the current climate change framework could be addressed.

Part III: The building blocks of forest financing: the medium-term approach

Part II described how a global cap and trade system that takes account of all sectors, including forests, would perform best as a framework for tackling climate change in the long term. However, a global climate change deal should take account of the different stages of development of developing countries. The global carbon market will need to evolve, with national and regional emissions trading schemes growing and merging over time. Furthermore, many developing countries may prefer to participate in a more incentive-based scheme in the near term.

Part III looks at the importance and key elements of a well-designed approach to transition over the medium term. First, Chapter 8 sets out the types of transition path to a global cap and trade system. Underpinning a successful transition path is a three-stage process: short-term, medium-term and long-term goal. The most effective transition path to global cap and trade is likely to be an incentive-based approach with increasing access to regional and national emissions trading schemes, while drawing on additional finance from other sources as carbon markets grow over time.

The next four chapters describe the four key building blocks which will be required if forest nations are to benefit from financial flows from developed countries while ensuring that their forest policies genuinely reduce carbon emissions.

Chapter 9 examines the types of incentive-based targets that are most effective in reducing forest emissions.

Chapter 10 goes on to review the technology, capacity building and costs required to ensure that reductions of forest emissions from the baseline are measured and monitored robustly.

Chapter 11 analyses the challenges of linking the forest sector of developing countries to emerging national and regional carbon markets, including the required scale of finance and impacts on carbon markets of forest credits. The chapter also provides an analysis of the funding gap that exists in the medium term as carbon markets mature and of the potential sources of funding that could fill the gap.

Finally, Chapter 12 examines the importance of good governance, including land tenure and policy incentives, as well as the different financial distribution mechanisms that forest nations may wish to use to ensure that finance reaches the most appropriate regions, communities, individuals and programmes efficiently and transparently.

8. Transition to a long-term framework

Key messages

The transition to a long-term goal of global cap and trade will need to meet the development needs of countries at different levels of development, particularly the poorest.

If the transition path is poorly designed, the long-term goal may not be reached or may be delayed. A smooth transition path is also important for building confidence in the system.

The most effective transition path to global cap and trade is likely to be an incentive-based approach with increasing finance from regional and national emissions trading schemes while drawing on additional finance from other sources while carbon markets grow over time.

Underpinning a successful transition path is a three-stage process covering the short term, medium term and the long-term goal.

In the short term, the key objectives should be capacity building and filling the funding gap. Capacity building and demonstration activities will be needed to build confidence and ensure that mechanisms and institutions are fit for purpose. A combination of public and private sector finance could be used for pump-priming credit mechanisms ahead of access to carbon market finance.

In the medium term, international and national systems should begin to provide access to emissions trading schemes for forest nations. Additional funding from the public and private sectors could be reduced as carbon market finance increases over time. Four building blocks are key:

- transitional arrangements for targets, or reference levels, that are no lose or limited liability for developing countries;
- capacity building for robust measuring and monitoring of forest emissions;
- a well designed system for linking forest credits to national and regional carbon markets that maintains financial stability while drawing on additional sources of public and private finance;
- strong governance.

In the long term the goal should be full inclusion in a global carbon market.

8.1 Introduction

The transition to a long-term goal of global cap and trade will need to meet the development needs of countries at different levels of development, particularly the poorest.

If the transition path is poorly designed, the long-term goal may not be reached or may be delayed. A smooth transition path is also important for building confidence in the system.

While a long-term framework for an emissions reduction system is essential if the international community is to stabilise global temperatures at around 2°C increase (see Introduction), the transition path towards that goal is equally important. If the transition path is poorly designed, the long-term goal may not be reached or may be delayed. Given the urgency of the challenge of climate change and the rapid increase in industrial emissions that the world is currently experiencing, any delay in tackling the challenge will substantially increase the risks from climate change.[1] A smooth transition path is also important for building the credibility of the system. Testing approaches to credit transfers for emissions reductions through demonstration activities at local, regional and national levels will be an important part of the process to build confidence and ensure that mechanisms and institutions are fit for purpose.

Part II of this Review concluded that a variety of systems exists for achieving reductions in deforestation as part of an overall global framework, which include public and private finance as a source of funding emissions reductions. Many of these systems are potentially valuable tools in tackling climate change and could operate in parallel. In the long term, a cap and trade framework seems best to meet the criteria of effectiveness, efficiency and equity and should play an important role in reducing emissions, including those from forestry. At the same time, implementation of the current international framework, while building credibility in an international system for emissions reductions, has significant limitations for meeting a global emissions target. In this chapter, we briefly review three options for moving from the current system towards long-term global carbon trading:

- moving immediately to a cap and trade system for all countries;
- accessing international finance solely from outside the global carbon market;
- accessing finance under incentive-based schemes from a combination of carbon markets (regional and national emissions trading schemes) and other sources while carbon markets grow smoothly over time.

Of these options, this Review argues that the third system, a combination of carbon market finance and additional public and private finance, will be most effective as a short to medium term transition to global cap and trade. The building blocks of this transition are set out in following chapters.

---

1   Stern (2007)

8.2 Types of transition path

The most effective transition path to global cap and trade is likely to be an incentive-based approach with increasing finance from regional and national emissions trading schemes while drawing on additional finance from other sources while carbon markets grow over time.

One option for transition to the long-term framework is to move immediately to a cap and trade system for all countries. This would have several advantages, including potentially large scale finance in a permanent system of emissions reductions. However, in practice it poses a number of challenges. Many developing countries are not in a position to face the liabilities associated with having a stringent binding cap. Many would argue that the developed world, being largely responsible for industrial emissions over the last 150 years, should shoulder a large responsibility for global emissions reductions. Meanwhile, it is important that emissions reductions for tackling climate change should not impede the development of poorer countries. Consequently, a transitional system in the medium term may need to include limited liability targets for many developing countries that act as part of an incentive-based scheme to reduce emissions (see Box 8.1). Lack of capacity for robust measuring and monitoring and governance will also be a challenge for many forest nations, and building capacity to a level sufficient for forest nations to participate in an emissions reduction mechanism will take time to achieve (see Chapter 13).

Given that a long-term framework of cap and trade will take some time to develop, another option for addressing the urgent challenge of deforestation in the short to medium term is to rely solely on sources of funding other than carbon markets. As set out in Chapter 6, while funding sources additional to carbon markets will be important to fill the funding gap in the near term, relying entirely on these sources would be very unlikely to deliver the scale of finance required to make a significant impact on emissions from the forest sector. Even assuming that the level of funding necessary will be lower in the early stages, as developing countries build the capacity to absorb large financial flows, the funding required will be substantial (see Chapter 13 for a more detailed discussion of funding sources).

A third option is to access finance from regional and national emissions trading schemes in the medium term under a more incentive-based scheme using baselines, or reference levels, while accessing additional finance from other sources as carbon markets grow smoothly over time (see Box 8.1). If developing countries are to have access to the substantial financial flows from a global cap and trade system in the future, a smooth and steady transition will be needed, with comprehensive preparation of international and national systems. As access to finance from carbon markets grows over the medium term, the remaining funding gap in the interim could be filled from additional sources from both the public and private sectors.

As part of an incentive-based approach for non-Annex I countries, some have argued for a middle way between commitment-free participation and the adoption of new binding commitments.[2] Countries wishing to graduate from project-based CDM to large-scale financial mechanisms would adopt targets, or reference levels, to incentivise abatement opportunities in the participating countries' economies (or specific sectors). Any emissions reduc-

---

2 Eg Frankel (2007)

tions below this target could be sold on the carbon market at the prevailing carbon price. As long as the real cost of the abatement were lower than the prevailing carbon price in the market, non-Annex I regions could more than cover the incremental cost of the abatement and would make profits. National baselines should reduce concerns over leakage, allowing Annex I countries to adopt more ambitious emissions reduction targets. The use of reference levels by non-Annex I countries would also signal their intention to mainstream climate change mitigation policies into national growth and development strategies.

Under one proposal, the Sao Paulo proposal that was presented to the UNFCCC negotiators in 2006, a new deal would offer a ladder for developing countries to graduate at their own pace, rather than an arbitrary timetable. All countries would continue to have access to baseline-credit mechanisms, but only while the volume of their abatement exports remained below a given share of the international abatement market. Beyond this threshold, the country would adopt some form of constraint in order to continue trading (see Chapter 9).

> **Box 8.1: An incentive-based forest scheme as part of a cap and trade system**
>
> A global climate change deal should take account of the different stages of economic development. Many developing countries are not in a position to face the liabilities associated with having a stringent binding cap below their expected business as usual emissions. Some countries may wish to have binding caps, enabling them to receive international allowances to the level of their target and sell any surplus allowances through reducing forest emissions. Furthermore, it is likely that part of the action plans that developing countries will need to adopt as part of an international effort to tackle climate change will involve a willingness to discuss binding caps for middle-income countries by 2020.[3] However, for many forest nations, a transitional system in the medium term may need to include no lose, or limited liability, targets that act as part of an incentive-based scheme to reduce emissions. Central to any system is that emissions reductions for tackling climate change should not impede the development of poorer countries.
>
> Many developing countries may be more ready to participate in the near term in a national-level baseline-credit system with limited liabilities. Under a baseline-credit system, credits can be earned at the end of a crediting period on the basis of whether and by how much emissions fall below a baseline or reference level. This provides an incentive for forest nations to reduce emissions without the potentially high penalties associated with a stringent binding cap. The credits produced from forest emissions reductions, and reductions from other sectors, can be traded with Annex I emitters who are subject to a binding cap, and so provide demand for international credits.
>
> In a market for emissions credits, demand for credits is generated by the gap between baseline emissions (eg historical emissions or other proxy for business as usual emissions) and the cap set in the emissions market. The supply of credits is determined by the cost of abating emissions. The credit supply curve is known as the Marginal Abatement Cost (MAC) curve.

---

3   Stern (2008)

The Figure below illustrates how the price of an emissions credit arises, and consequently the overall cost of abatement. The triangle under the MAC curve represents the resource cost of abatement, while the price of emissions credits is the point at which the abatement supply curve and the demand curve intersect, P*.

Further details of setting baselines, or reference levels, and linking credits to carbon markets are set out in Part III of this Review.

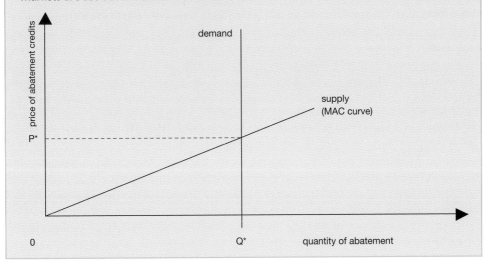

For the reasons set out above, and in Chapters 5, 6 and 14, a smooth transition with a combination of funding from carbon markets and additional public and private sector finance seems to be the most effective approach for reducing forest emissions in the short to medium term. It can also be argued that this approach is suited to other emitting sectors in developing countries.

8.3 A three-stage transition process: short, medium and long term

Underpinning a successful transition path is a three stage process covering the short term, medium term and the long-term goal.

In the short term, the key objectives should be capacity building and filling the funding gap. Capacity building and demonstration activities will be needed to build confidence and ensure that mechanisms and institutions are fit for purpose. A combination of public and private sector finance could be used for pump-priming credit mechanisms ahead of access to carbon market finance.

In the medium term, international and national systems should begin to provide access to emissions trading schemes for forest nations. Additional funding from the public and private sectors could be reduced as market access increases over time. Four building blocks are key:

- transitional arrangements for targets, or reference levels, that are no lose or limited

liability for developing countries;
- capacity building for robust measuring and monitoring of forest emissions;
- a well designed system for linking forest credits to national and regional carbon markets that maintains financial stability while drawing on additional sources of public and private finance;
- strong governance.

In the long term the goal should be full inclusion in a global carbon market.

Underpinning a successful transition path is a three stage process: short term, medium term and long term (see Figure 8.1). In the short term, over the next four to five years, the key objectives should be capacity building and filling the funding gap. Many countries will need to undertake a range of preparatory work, reforms and capacity strengthening measures before they can participate fully in a forest credit mechanism. International institutions will also need to be reformed. This capacity building should include approaches to financial transfers for emissions reductions through demonstration activities at local, regional and national levels to build confidence and ensure that mechanisms and institutions are fit for purpose. As well as capacity building in preparation for carbon market access, other sources of international funding should be provided to meet the funding gap. A combination of public and private sector finance could be used for pump-priming credit mechanisms ahead of access to carbon market finance. Short term capacity building, and short to medium term finance to meet the funding gap, are discussed in more detail in Chapter 14.

Figure 8.1: Transition path to a long-term framework for global carbon
Funding

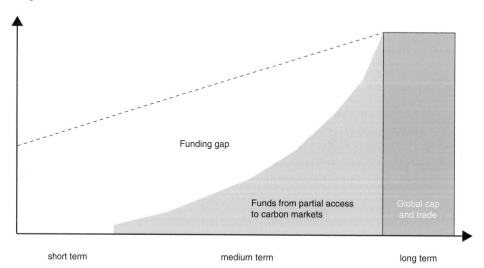

emissions reductions

In the medium term, beyond 2012, international and national systems should begin to provide carbon market access for forest nations which can demonstrate sufficient levels of capacity. Market access should grow over the medium term. While a smooth transition with funding from combined sources is an effective approach in the short to medium term, further challenges present themselves. In particular, supply and demand of carbon credits need to be managed carefully, and the introduction of international credits, including those from the forest sector, should be implemented under a well designed system.

Four building blocks are key to a successful transition over the medium term. First, transitional arrangements may be needed for targets, or reference levels, that reduce the liability for developing countries if emissions reductions increase beyond governments' control, while providing the incentive of finance flows for delivering real emissions reductions. Second, capacity building will be required in many cases at national and international levels for robust measuring and monitoring of forest emissions. Third, a well designed system for linking forest credits to emissions trading schemes is essential to create sufficient demand and maintain financial stability while drawing on additional sources of public and private finance. Finally, governance at national and international levels will be important, requiring some major reforms in some areas of policy and institutions at all levels. Chapters 9 to 12 in Part III of this Review examine each of these four medium-term building blocks in turn.

8.4 Conclusion

The transition path towards a long-term goal of global cap and trade will be critical to the success of an international emissions reduction system. The most effective transition path is likely to be an incentive-based approach with increasing access to regional and national emissions trading schemes while drawing on additional finance from other sources as carbon markets grow over time. This combination is important for raising sufficient international finance to tackle emissions from the forest sector through REDD and ARR. Nonetheless, it also raises challenges of its own, not least the coordination of finance to ensure that funds are channelled effectively and efficiently, as well as being distributed equitably across and within forest nations.

Underpinning a successful transition path is a three stage process: short term, medium term and long-term goal. In the short term, the key objectives should be capacity building and filling the funding gap. These are examined in detail in Chapter 14.

In the medium term, international and national systems should begin to provide carbon market access for forest nations, with market access growing over time. Four building blocks will be necessary if the transition process is to be successful. First, transitional arrangements for targets - baselines or reference levels - that are no lose or limited liability for developing countries will probably be necessary. Second, capacity building for robust measuring and monitoring of forest emissions will be essential. Third, a well designed system for linking forest credits to national and regional carbon markets will be required that maintains financial stability while drawing on additional sources of public and private finance. And finally, the system will need to be underpinned by strong governance at all levels. The following four chapters examine each of these building blocks in turn.

9. Effective targets for reducing forest emissions

Key messages

Effective targets for reducing forest emissions need to:

- minimise leakage (reductions in deforestation in one area leading to increases in another);
- ensure real reductions compared with business as usual (additionality);
- incentivise action to retain or enhance standing forests.

National baselines or reference levels should be used to prevent intra-national leakage. A baseline-credit system for non-Annex I countries could initially generate credits for emissions reductions in forest emissions on a no-lose or limited liability basis. This could be undertaken under a new mechanism or within a radically reformed Clean Development Mechanism.

To ensure that emissions reductions are additional, incentives should be linked to global business as usual emissions.

Baselines that take account of the global average deforestation rate can incentivise action to retain or enhance standing forests. Credits for avoided deforestation would represent payment for a global service, especially as successful action in high-deforesting nations may increase pressure to deforest in nations where deforestation rates are currently low.

In order to meet all the above criteria, baselines should take account of a country's historical emissions rate and the global average deforestation rate. This will ensure that emissions reductions in the global forest sector are additional while acting against international leakage by being inclusive.

Baselines should change over time by means of a renegotiation linked to an indicative trajectory which would help ensure additionality and facilitate reaching agreement.

9.1 Introduction

Effective targets for reducing forest emissions need to:

- minimise leakage (reductions in deforestation in one area leading to increases in another);
- ensure real reductions compared with business as usual (additionality);
- incentivise action to retain or enhance standing forests.

Effective targets for reducing net emissions from deforestation (taking account of deforestation, degradation and ARR) are a prerequisite for successfully tackling the impacts of forests on climate change. An effective, performance-based target requires a baseline or reference level which minimises leakage, ensures that emissions reductions are additional and incentivises action to protect or enhance standing forests. This chapter examines the most suitable level and type of baseline to be most effective, efficient and equitable. The following chapters go on to examine the importance of robust measuring and monitoring, and linking generated forest credits to trading systems.

9.2 Baseline level

National baselines or reference levels should be used to prevent intra-national leakage. A baseline-credit system for non-Annex I countries could initially generate credits for reductions in forest emissions on a no-lose or limited liability basis. This could be undertaken under a new mechanism or within a radically reformed Clean Development Mechanism.

The level against which performance is assessed affects the effectiveness, efficiency and equity of a baseline-credit system, particularly the level of leakage (reductions in deforestation in one area leading to increases in another).

9.2.1 Baseline level options

Sub-national-level baselines

Baselines can be established for individual projects, as is the case for afforestation and reforestation (A/R) projects under the Clean Development Mechanism (CDM). In this case, the target area is treated as an independent unit. A business as usual reference level is established and credits are awarded when emissions are below this level. An important advantage of sub-national-level baselines is that the outcome of the project is free of uncertainties related to deforestation levels outside the project area. This feature is important where uncertainties related to governance issues could discourage investors from projects if their outcome is influenced by what happens outside the project area.

    This independence, however, is also a weakness of project level accounting. The forestry sector is widely considered particularly vulnerable to leakage. In this context, sub-national-level accounting might lead to a situation where successful individual projects are rewarded but their gains are totally or partially offset by increased emissions from deforestation in areas outside the area covered by the project. The risk of leakage with project-based approaches is considered to be much greater for deforestation projects than for A/R projects. This is a major reason why A/R projects have been included under the project-based CDM for the first commitment period under the Kyoto Protocol, while deforestation projects have been excluded.

National-level baselines
Baselines could be established at the country-wide level. In this case, credits would be awarded when the total emissions from the forest sector in the country falls below a national reference level. An important advantage of this level of accounting is that it captures intra-national leakage, as incentives are connected to the net reduction in emissions at the national level. Another important advantage of national-level accounting is that it is likely to lead to a greater involvement of national governments. And it is widely accepted that some of the most important underlying causes of deforestation are decided or heavily influenced by national-level policies.[1] The promise of sustained incentives might shift some of these long-term development policies towards a more sustainable pathway. Affecting these policies is a key factor for the success of a scheme to reduce emissions from deforestation (see Chapter 4).

National-level accounting does not solve all the issues. Potential for international leakage remains. It has been suggested[2] that the total incentive to be paid to forest nations could be calculated at a global level and subsequently divided among individual nations according to their relative performance. Although this option could eliminate the threat of leakage at all levels, it would increase the uncertainties to each individual forest nation, which could in turn lead to limited mitigation efforts. The most practical way of avoiding international leakage is probably to design an incentive system that is acceptable to a broad range of countries.

9.2.2 Advantages of national-level baselines

National-level baselines are the most appropriate for an international agreement between nations (see Table 9.1). Although not a perfect solution, they deal with the serious threat of intra-national leakage and are likely to encourage the participation of national governments. As many of the key drivers of deforestation are strongly linked to national-level policies, involvement of national governments is important for the success of a scheme to reduce emissions from deforestation. The adoption of national-level baselines respects the sovereignty of nations and gives them flexibility to decide what sort of internal distribution of incentives is best suited to them. Due to considerations such as these, the Bali Decision on deforestation in developing countries (in indicative guidance for demonstration activities), states that: *'emission reductions from national demonstration activities should be assessed on the basis of national emissions from deforestation and forest degradation.'*[3]

It is important to bear in mind, however, that incentives need to reach local actors in the deforestation process. The risk of marginalisation of local communities, sometimes associated with centralised schemes, also needs to be addressed. Furthermore, adopting national-level baselines in an international agreement does not preclude countries from exploring sub-national-level activities within its borders. Many countries may wish to adopt a sub-national scheme of incentives partially connected to the national scheme. These issues are dealt with in more detail in Chapter 12.

---

1 Geist and Lambin (2002)
2 Strassburg et al (2007); Mollicone et al (2007)
3 Decision 2/CP.13: Reducing emissions from deforestation in developing countries: approaches to stimulate action (2007)

Given that under the current international system, REDD is not included under the CDM while A/R activities are covered only at a sub-national (project) level, moving to a national approach will require substantial reforms. One option is the creation of a new mechanism outside the CDM for setting baselines, monitoring emissions reductions and accreditation. Alternatively, the CDM would need substantial reforms for scaling up projects to the national level. Some progress has been made in scaling up from projects to programmes at the sub-national level. Parties agreed at Montreal in 2005 that projects could be undertaken under a Programme of Activities (PoA) to be considered by the CDM board.[4] As part of a study commissioned by this Review on scaling up forestry and other land-use mitigation activities in non-Annex I countries, the majority of CDM project developers interviewed were reluctant to be pioneers in undertaking a PoA due to the unknown cost, time and complexity of the additional processes involved.[5] In addition to these potential barriers to participation, the international community will need to determine whether the CDM, originally designed as a project-level mechanism, can be reformed sufficiently rapidly into a national-level mechanism with different objectives and procedures.

Table 9.1: Assessment of national versus sub-national baseline levels

Level	Effectiveness	Efficiency	Equity
Sub-national *Each project is treated as an independent unit*	(+) Safer for individual project investors as payoff is independent from what happens outside. (-) Highly vulnerable to leakage. (-) Might not get the involvement of the national governments/ influence national policies.	(-) Higher transaction costs. (-) Project area approach may be less cost-effective than a national policy approach.	(-) Marginalised groups may have less access to the market.

---

4 Decision CMP1: Further guidance relating to the clean development mechanism (2005)
5 Baalman and Schlamadinger (2008)

Level	Effectiveness	Efficiency	Equity
National *Refers to the net national emissions from deforestation in a given country*	(+) Captures intra-national leakage. (-) Still vulnerable to international leakage. (+) Government policy levers are activated (but incentives need to reach local actors) (+) Possible influence on long-term development policies towards sustainable development (-) Might discourage investors from individual projects, especially when governance challenges are present.	(+) Possible economies of scale on transaction costs. (+) Changing/enforcing policies/laws could be very cost effective.	(+) Might promote more equitable distribution of benefits within country. (-) Risk of infringements of indigenous peoples' rights.

9.3 Determining the baseline

To ensure that emissions reductions are additional, incentives should be linked to global business as usual emissions.

Baselines that take account of the global average deforestation rate can incentivise action to retain or enhance standing forests. Credits for avoided deforestation would represent payment for a global service, especially as successful action in high-deforesting nations may increase pressure to deforest in nations where deforestation rates are currently low.

In order to meet all the above criteria, baselines should take account of a country's historical emissions rate and the global average deforestation rate. This will ensure that emissions reductions in the global forest sector are additional, while acting against international leakage by being inclusive.

Credits would be awarded for emissions reductions below a baseline, or reference level, at the end of an agreed crediting period, which is the time over which an agreed baseline remains valid for accounting purposes (see Figure 9.1). Baselines are often thought of in terms of historical emissions, but the level need not be purely historical, particularly when countries with a track record of low rates of deforestation are considered. This section sets out the different options and appraises them, particularly with regard to:

- the need to ensure real reductions compared with business as usual (additionality);
- international leakage (reductions in emissions in one country causing increases in emissions in another);
- the incentive to retain or enhance standing forests.

References to emissions from deforestation also include emissions from degradation, unless stated otherwise.

Figure 9.1: Illustration of a baseline-credit system

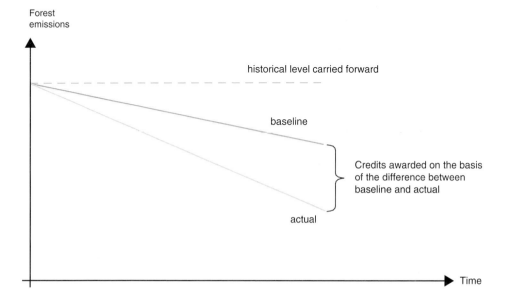

9.3.1 Types of baseline

Historical baselines

The first scientific[6] and national[7] proposals for incentive mechanisms to reduce emissions from deforestation suggested that reference levels should be set based on historical emissions. The baseline could be set based on emissions that occurred before a particular historical date, such as the start of the current discussions in the UNFCCC in December 2005. It has been proposed that a crediting period of five to ten years should be used, to help smooth inter-annual variations in deforestation rates. The general formula for a historical baseline is described in Box 9.1.

---

6 Santilli et al (2005)
7 UNFCCC (2006)

Box 9.1: Formulae for crediting against historical and stock/average baselines

Historical baseline

Under a historical baseline, emissions reductions are calculated as the difference between actual net emissions from deforestation, degradation and forestation at the end of a time period (Et) and the actual emissions at the beginning of the time period (PE). The financial reward for the reduction would also depend on the price per tonne of carbon ($ per t CO_2e). This is summarised in the following equation (note that emissions are net emissions from deforestation, degradation and forestation, which can be positive, zero or negative).

Equation (1): $I_t = (PE - Et) \times (\$ \text{ per t } CO_2e)$

where I_t = a country's incentive in year t; PE = past emissions as annual average emissions in a given period; and Et = emissions in year t.

Stock/average baseline

Under a stock/average baseline, emissions reductions are calculated as the difference between actual net emissions at the end of the time period (Et) and the expected emissions from that country if it were to deforest at the global average. This is summarised in the following equation:

Equation (2): $I2 = EE - Et$

where EE is the expected emissions from that country (the product of its forest/carbon stock and the average global deforestation/emissions rate).

An advantage of this approach is that it offers greater incentives to countries that have experienced high deforestation in the recent past. This may lead not only to relatively faster reductions in deforestation in the short term, but also increases short-term additionality. However, the historical approach has some drawbacks (see Table 9.2). On the equity side, it might not be desirable for countries that have been protecting their forests to be penalised by their 'good' past behaviour. In addition, the reduction of deforestation rates in high-deforesting countries would probably lead to increased pressure for conversion of forests elsewhere, particularly for agricultural expansion. The potential for international leakage towards those countries that would be receiving low or no incentives would increase. That could lead to a situation where high-deforesting countries are rewarded by reducing their rates but increased deforestation in other countries offsets the mitigation benefits.

Stock/average emissions baselines

Some have argued[8] that incentives should be directly connected to forest area or forest carbon stock, regardless of countries' past deforestation rates (see Box 9.1 for a general

---

8 Strassburg et al (2007); TCG (2008)

formula for a stock/average emissions baseline). The goal would be to maximise the emissions reduction scheme's coverage by increasing the value of standing forests throughout the developing world. This broader coverage would minimise the threat of international leakage.

This approach requires careful attention to additionality. The potential for rewarding more reductions than would occur in the absence of the scheme is high, since the annual deforestation rate is relatively low compared with total forest stock (around 0.2 per cent[9]). These rewards are often referred to as hot air.

An alternative proposal[10] assumes that all carbon contained in legally, physically and economically accessible areas in the developing world would be emitted within the next 50 years. Each country would calculate its total accessible carbon and a baseline would then be constructed assuming that all the accessible carbon would be emitted at a constant rate over the next 50 years.

Another proposal[11] incorporates a direct link with forest/carbon stock into the standard UNFCCC baseline concept. The authors argue that it is fair to assume that over the long run all developing countries would deforest at the average global rate. A baseline would be estimated based on the global average emissions rate and each country's carbon content. Under this approach, the sum of all countries' baselines equals the global baseline. As a consequence, regardless of the level of additionality for each country, additionality at the global level would be guaranteed and the scheme would be free of hot air. However, there are several challenges to making this approach workable (see Table 9.2).

Projected baselines
A projected, or business as usual, baseline anticipates future emissions from deforestation, usually based on past deforestation and projections for key social, economic, political and technological variables. Some regional estimates have been published based on models that incorporate some or all of these variables.

In theory, projected baselines are the perfect reference levels. If it were possible to predict when and where deforestation would occur without an incentive scheme, additionality would be maximised and incentives would be offered at the appropriate level to all forest countries, minimising the risk of leakage.

In practice, the mix of drivers behind deforestation are complex and difficult to quantify accurately (see Chapter 3), while the future behaviour of key variables is uncertain. In addition, projected baselines provide perverse incentives to countries which would benefit from publishing future plans for major deforestation activities simply to receive more finance for curtailing them. Consequently, it is unlikely that a projected baseline based on model results would be an acceptable target for reducing emissions. However, projection models and the underlying understanding of past trends are likely to be used in negotiations to agree country baselines and will be of great importance in the design of national or local policies to tackle deforestation.

Combined (historical and average) baselines
A fourth type of baseline combines historical and average baseline levels, maximising the strengths and minimising the weaknesses of each.

---

9   FAO (2006)
10 TCG (2008)
11 Strassburg et al (2007)

One approach would be to use historical emissions rates for high-deforesting countries while offering low-deforesting countries a fraction of the global average deforestation rate as their baseline rate.[12] In this way, high-deforesting countries would receive incentives to join the scheme in the short term and low-deforesting countries would receive greater incentives than under a scheme based solely on historical rates.

The total incentive can be calculated on the basis of global performance and then allocated to forest countries according to their relative performance. This step would help ensure global additionality and take into account international leakage. On the other hand, it would make the incentives received by each forest country partially connected to the behaviour of all other forest countries. In addition to being politically sensitive, this would increase the uncertainties for each forest country. This approach would continue to give the highest incentives to countries with the highest deforestation rates, which might be viewed as inequitable.

Another approach would offer forest countries incentives for improved performance compared with both historical and average rates (see Box 9.2). The first incentive is based on a historical baseline (in a similar way to the pure historical baseline approach). The second incentive is based on a global average baseline. A country would be rewarded for emitting less than the global average. This acts as a safeguard against international leakage by giving forest nations with standing forests and a record of avoided deforestation an incentive not to revert to deforestation (this is similar to the average baseline approach).

All forest countries would receive a combined incentive with the historical and average components weighted together. An interesting feature is the way in which these incentives could be combined. It has been shown that by making the weight of each incentive variable, a scheme based on these combined incentives could be sufficiently comprehensive to include all countries in a single formula and flexible enough to combine short-term realities with long-term sustainable goals.[13]

> **Box 9.2: Combined (historical and average) baseline**
>
> A combined historical and average baseline can maximise the strengths and minimise the weaknesses of the historical and average baselines when used independently. The following formula is one example of this combination approach.[14] The first incentive would reward a country for reducing emissions in relation to a baseline period (ie against historical emissions):
>
> Equation (1): $\quad I1 = PE - Et$
>
> where PE is the past emissions from that country in an agreed period and Et is its emissions in year t.

---

12 Mollicone et al (2007)
13 Strassburg et al (2008)
14 Strassburg et al (2008)

> The second incentive would reward a country for maintaining emissions below the global average rate:
>
> Equation (2): $I2 = EE - Et$
>
> where EE is the expected emissions from that country (the product of its forest/carbon stock and the average global deforestation/emissions rate).
>
> Under the combined baseline approach, all forest countries would receive both incentives at the same time. The relative weight of each incentive is adjustable. The factor α weights the incentives between historical and average/stock incentives. So the mechanism baseline formula would be:
>
> Equation (3): $CI = [\alpha (I1) + (1 - \alpha)(I2)] \times (\$ \text{ per t } CO_2 e)$
>
> where α ranges from 0 to 1. Countries with high historical rates of deforestation receive incentives to reduce forest emissions from the historical component of the equation. At the same time, countries with standing forests and a track record of avoided deforestation receive incentives to keep deforestation rates low, zero or negative (eg if rates of ARR are high), rewarding good forest policies and reducing the risk of international leakage of deforestation to these countries. By offering incentives to countries with both high and low deforestation rates, such an approach could be sufficiently comprehensive to minimise international leakage.
>
> Another interesting feature of this model is that, if the weighting factor α is the same across all countries, the weighted sum of all countries' baseline equal the global baseline. As a result, even if additionality is not met in a specific country, global-level additionality would still be guaranteed. In this way, the mechanism would not generate hot air compared with the deforestation rate at the time of the agreement. If it were decided that the business as usual global deforestation rates should go down after a certain period of time, these terms could be modified proportionately and additionality could be maintained.

9.3.2 Assessment of different baseline types

The Bali Decision on deforestation in developing countries states that: *reductions in emissions or increases resulting from the demonstration activity should be based on historical emissions, taking into account national circumstances.*[15] A combined baseline need not be inconsistent with this goal, as emissions would still be estimated against a historical reference. At the same time, adding the component based on global average emissions may help reach agreement.

Baselines based on stock or average emissions rates would be likely to have a comprehensive coverage over the long term, as the scheme would be directly related to the remaining forest area or carbon stock. This would offer incentives to countries in all stages of the deforestation process. However, given that deforestation rates in some countries are currently substantially higher than the average global rate, it would offer these high-deforesting countries little incentive to try to reduce rates below the average.

---

15 UNFCCC Decision 2/CP.13

Consequently, stock/average baselines are unlikely to be negotiable and even if put in place would probably result in high-deforesting countries simply allowing high rates to continue (see Table 9.2).

Table 9.2: Assessment of four types of baseline

Type	Effectiveness	Equity
1. Historical *Based solely on past emissions from each country*	(+) Higher incentives to countries currently with higher deforestation rates. (+) Suitable for short-term reductions. (-) Little incentives to countries with low, zero or negative deforestation rates in the recent past. (-) Vulnerable to international leakage to countries that would receive low incentives. (-) Not suitable as long-term solution.	(-) Offers smaller rewards to countries that have been protecting their forests.
2. Stock/average *Based on current forest/carbon stock of each country and possibly a global average deforestation/emissions rate*	(+) Incentives offered to countries in all stages of the deforestation curve might minimise risk of leakage. (+) Directly connected to remaining forest area/carbon stock. (+) Using average global rates for all countries eliminates risk of hot air. (-) But a pure carbon stock approach would produce substantial hot air. (-) Little incentive for high-deforesting countries to join in the short term.	(+) Does not penalise or reward countries for behaviour prior to the establishment of the scheme.
3. Projected *Based on past deforestation and estimates of future deforestation drivers and key social, economic, political and technological variables*	(+) If possible to estimate, offers highest level of additionality. (+) If properly estimated, should offer appropriate incentives to countries in different stages of the deforestation process, minimising risk of leakage. (-) Sensitive to model assumptions about deforestation drivers and future behaviour of key economic, social, political and technological variables. (-) Individual model results too uncertain to be agreed as the base of an international agreement. (-) Can create perverse incentives.	(+) Socio-economic circumstances would be reflected. (+) Understanding needed to develop models would be relevant to negotiated outcomes.

Type	Effectiveness	Equity
4. Combined (historical and average) *A combination of historical and average baselines*	(+) Could be sufficiently comprehensive to include countries at all stages of the deforestation process. (+) Minimises risk of international leakage. (+) Adjustable across countries, if required by specific national circumstances. (+) Adjustable over time, conciliating short- and long-term incentives needs. (+) Based on simple and transparent available data. (+) If properly calibrated, can guarantee additionality among participating countries. (=) Requires proper calibration.	(+) Can converge to the equitable distribution of average baselines over time. (-) Offers higher rewards to high-deforesting countries, although to a lesser extent than purely historical baselines.

A combined baseline has the potential to be sufficiently comprehensive to attract countries at all stages of the deforestation process over both the short and long term. Countries with high historical rates of deforestation receive strong and realistic incentives to reduce forest emissions. At the same time, countries with standing forests and a track record of avoided deforestation would receive incentives to keep deforestation rates low, zero or negative (if, for example, rates of ARR are high). This rewards countries with a history of responsible forest policies while reducing the risk of international leakage of deforestation to these countries. Achieving the appropriate balance between additionality and avoided international leakage is key. A baseline with a flexible combination of historical rates and incentives for countries with low deforestation or net sequestration could therefore be the most equitable and effective type of baseline for a scheme of incentives to reduce emissions from deforestation.

Modelling commissioned for this Review examined the effectiveness and equity of different types of baseline.[16] The results showed that the combined baseline performed best in terms of incentives for a wide range of forest nations, while providing a balance between additionality and leakage (see Table 9.3). Under a historical baseline ($\alpha = 1$), the results are highly inequitable, with high-deforesting countries gaining the most and low-deforesting countries gaining the least as predicted. The financial gains to some forest nations with very high levels of deforestation are over twice that of the average in the model. At the same time, countries with a track record of avoided deforestation received no incentives at all. This would put significant pressure on these countries to start deforesting, leading to international leakage.

Under a stock/average baseline ($\alpha = 0$), low-deforesting countries and countries with no deforestation gain 45 per cent of the total incentives in the model. This is despite the fact that deforestation rates are not lowered substantially in these countries relative to historical emissions. Furthermore, as set out above, it is unlikely that high-deforesting

---

16 Strassburg (unpublished)

countries would be able to reduce deforestation rates sufficiently rapidly to receive any incentive in the short term.

In contrast, under a combined baseline model (Box 9.2), forest nations with high, low and zero net deforestation rates all gain. This is only one of various models for a combined baseline, and the results should be seen as indicative. Any baseline will need to be carefully negotiated by the international community to ensure the right balance of additionality, prevention of international leakage and equity for developing countries.

Table 9.3: Financial incentives for forest nations under different types of baselines

Country (level of deforestation)	Historical deforestation (%)	Financial incentives (% of total)			
		Historical baseline	Combination baseline	Combination baseline	Stock/average baseline
		$\alpha = 1$	$\alpha = 0.9$	$\alpha = 0.5$	$\alpha = 0$
Very High	1.1	44	41	30	16
High	0.7	28	27	23	19
Average	0.5	20	20	20	20
Low	0.2	8	9	15	22
Zero	0.0	0	2	12	23
Global	0.5	100	100	100	100

Note: See Box 9.2 for principles of the model, including levels of α. Model runs based on a carbon price of $30 per tonne. All countries in the model are assumed to have the same forest area, the same carbon stock, and reduce their deforestation rates from the historical level by the same proportion.

9.4 Baseline trajectories

Baselines should change over time by means of a renegotiation linked to an indicative trajectory which would help ensure additionality and facilitate reaching agreement.

Forest emissions baselines are unlikely to remain static over time. The principal reason is that business as usual forest emissions projections are not static. Many forecasts (though not all – see Chapter 2) project declining forest emissions over time as more forest area is lost to deforestation and degradation and consequently fewer opportunities exist to clear forest. Baselines need to move downwards as deforestation rates are reduced to ensure that credits awarded against them are additional. Furthermore, as developed countries address their historical liability for current emissions through deeper cuts and provision of funds to developing countries, the latter will be expected to take on greater climate change commitments themselves over time. This should include baselines that decrease over time at a higher rate than business as usual projections.

There are two principle means by which baselines can be adjusted: periodic negotiation or a more automatic adjustment according to a pre-determined trajectory. Negotiations provide more flexibility, particularly if country-specific circumstances need to be taken into account. On the other hand, a rules-based trajectory can provide clearer targets that are not subject to political interference. The best solution, and probably one that is most likely to be negotiable, is a system under which baselines change over time by means of a renegotiation linked to an indicative trajectory which would help ensure additionality (see Figure 9.2).

The indicative trajectory would need to be set according to a number of factors. These should include:

- a target for emissions reductions that contributes to a global stabilisation trajectory that minimises the risk of dangerous climate change;
- the impact on mitigation costs of starting to make deep cuts sooner rather than later;
- realistic incentives for forest nations given their current emissions levels and the time it would take to make the governance and policy changes necessary to produce significant emissions reductions.

Figure 9.2: Illustration of baseline trajectory over time

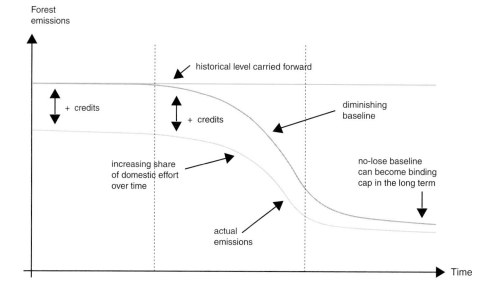

9.5 Conclusion

Effective targets are essential for a successful climate change framework that genuinely leads to strong reductions in carbon emissions. The forest sector is no different from other sectors in this regard. With the right type of baseline, or reference level, emissions reductions that take account of deforestation, degradation and ARR can be made effective and equitable.

A combined baseline has the potential to be sufficiently comprehensive to attract countries in all stages of the deforestation process over both the short and long term. Countries with high historical rates of deforestation receive strong and realistic incentives to reduce forest emissions. At the same time, countries with standing forests and a track record of avoided deforestation receive incentives to keep deforestation rates low, zero or negative (for example, if rates of ARR are high). This incentivises action to retain and enhance standing forests while reducing the risk of international leakage of deforestation to these countries. Getting the right balance between additionality and avoided international leakage is key. A baseline with a flexible combination of historical rates and incentives for afforesting and low-deforesting countries is therefore the most equitable and effective type of baseline for a scheme of incentives to reduce emissions from deforestation.

While the right type of baseline is key, it can only be as effective as the strength of national measuring and monitoring procedures. The next chapter sets out the importance of technology and capacity building to ensure that measuring and monitoring forest emissions at the national level is robust.

10. Measuring and monitoring emissions from forests

Key messages

Emissions reduction targets can only be monitored effectively if carbon emissions are estimated robustly and uncertainties are managed and quantified.

Robust measurements can be achieved using appropriate techniques and comprehensive and internationally consistent approaches. Forest emissions can be estimated with similar confidence to emissions estimates for other sectors.

Satellite technology, which offers transparent, consistent and rapid global coverage, can be used to map forests and identify their characteristics. This is then combined with research on the ground to calibrate satellite data and provide carbon density information.

Deforestation, afforestation and reforestation can be monitored effectively using satellite technology to assess changes in forest cover and land use. Monitoring forest degradation is more challenging, although recent developments in remote sensing, coupled with ground work and additional data such as proximity to roads, can be effective.

The IPCC has developed internationally agreed measures and guidance and these should be used to achieve consistency in monitoring and transparency in reporting. The IPCC Good Practice Guidance has resulted in major improvements in developed country estimates of agriculture, forestry and other land use. However, capacity building and national-level research will be needed in many other countries.

In countries where capacity is still developing, methodologies used may be less accurate. In these cases, conservative estimates of emissions and emissions reductions should be used.

Some of the poorest countries will not be able to build sufficient capacity in the short term. For these countries, similar methods described for national-level measuring and monitoring could be applied to sub-national programmes.

Review and verification of forest emissions are important to ensure that reductions are additional and long-lasting. Satellite images of changing forest cover provide a greater degree of transparency in monitoring forest emissions reductions than monitoring in other sectors.

> An estimated $50 million will be needed for a sample of 25 forest nations to set up national forest inventories, with a further $7-17 million needed for annual running costs. Technological advances, knowledge sharing and international cooperation can reduce these running costs over time.

10.1 The importance of robust measuring and monitoring

Emissions reduction targets can only be monitored effectively if carbon emissions are estimated robustly and uncertainties are managed and quantified.

Robust measurements can be achieved using appropriate techniques, and comprehensive and internationally consistent approaches. Forest emissions can be estimated with similar confidence to emissions estimates for other sectors.

Emissions reductions and targets are only as effective as the robustness with which they are calculated (see previous chapter for a discussion of targets). Emissions reductions can be underestimated or overestimated. Overestimating emissions reductions can lead to the appearance of targets having been met when this is not the case, with higher levels of emissions entering the atmosphere than reported. The risk of failure to meet global targets therefore increases if emissions reductions are not adequately measured, monitored and verified.

A robust system of measuring and monitoring requires that emission inventories are comprehensive and internationally consistent to enable verification of emissions reductions; and that uncertainties in measurement are quantified and managed. This chapter focuses on the factors influencing carbon stocks in forests; the process of measuring and monitoring forest stocks and emissions; and reducing the uncertainties in measurement through best practice technologies and techniques.

The capacity of forest nations to measure and monitor their forest emissions and sequestration varies greatly. Methodologies need to accommodate this variation to allow as many nations as possible to produce comparable outputs. The IPCC has developed internationally agreed Good Practice Guidance for compiling emissions inventories, and provided the framework for using remote sensing and ground based data to achieve consistency in monitoring and transparency in reporting. Progress towards this approach, and the costs and resources needed to build the necessary capacity to implement it, are presented in the second half of this chapter.

10.2 Measuring carbon stocks in forests

Satellite technology, which offers transparent, consistent and rapid global coverage, can be used to map forests and identify their characteristics. This is then combined with research on the ground to calibrate satellite data and provide carbon density information.

10.2.1 Factors influencing carbon stocks in forests

There is wide geographic variation in forest carbon stocks. A range of factors influence the quantity and distribution of these stocks, including climate, geology, tree maturity and natural disturbances.

- Climate plays a major role in determining carbon densities (see Chapter 2). The FAO's classification of ecological zones uses climate as the main criterion to categorise forests and the IPCC has collated estimates of above-ground carbon densities in these forest types (see Figure 10.1).
- Geology and soil type can exert a strong influence on the quantity and distribution of carbon stored in a forest. For example, the carbon stored in the active soil layer of tropical peat forests in Borneo can be four times higher than that in tropical lowland forests in Bolivia.[1]
- As forests sequester carbon as they grow, the maturity of a forest affects its carbon density. This is particularly important when considering the carbon stocks of areas being forested and harvested. For example, a species of Pinus grown in tropical Africa holds on average nearly four times more above-ground carbon in forest plantations that are over 20 years old than in those less than 20 years old.[2]
- Finally, disturbances such as fire, floods and droughts can result in carbon losses. These can be human induced or naturally occurring. For example, intense fires are part of the natural disturbance regime in the southern boreal mixed wood forests of Quebec.[3] Carbon stocks of the forest fluctuate as a result of disturbances, and are higher before a fire than afterwards. Knowledge of the disturbance regime can therefore help calculate more robust, time-averaged forest carbon stocks.

---

1 Brown et al (2008)
2 IPCC (2003)
3 Bergeron and Harveu (1997)

Figure 10.1: FAO Ecological zones and IPCC above-ground carbon densities

Key – Forest type, criteria and above-ground carbon (tCO$_2$/ha)

10. Measuring and monitoring emissions from forests

	Tropical	Subtropical	Temperate	Boreal
Criteria	All months without frost, and in marine areas at a temperature > 18°C	≥ 8 months at a temperature ≥ 10°C	4-8 months at a temperature > 10°C	≤ 3 months at a temperature > 10°C
Mainly wet months	Rainforest: 550 (220 – 1247)	Humid forest: 403 (18 – 1027)	Oceanic forest: 330 (147 – 2200)	Coniferous forest: 92 (18 – 165)
	Moist deciduous: 330 (18 – 1027)		Continental forest: 220 (18 – 587)	
Mainly dry months	Dry forest: 238 (183 – 752)	Dry forest: 238 (183 – 752)	Steppe	Tundra woodland: 27 (6 -37)
	Shrubland: 128 (37 – 367)	Steppe: 128 (37 – 367)		
All months dry	Desert: Negligible	Desert: Negligible	Desert: Negligible	
High altitudes	Mountain system: 257 (73 – 660)	Mountain system: 257 (92 – 660)	Mountain system: 183 (37 – 1100)	Mountain system: 55 (22 – 92)

Source: based on IPCC (2006) and FAO (2001)

10.2.2 Techniques for measuring carbon stocks

Forest types with similar carbon densities and profiles can be identified by recognisable characteristics. Uncertainty over different carbon densities and profiles of forests can therefore be reduced by (i) mapping forests; (ii) stratifying forest areas into ecosystem types with similar characteristics and sampling accordingly; and (iii) estimating carbon stocks.

Mapping of forests can be undertaken from the ground and from above. Ground observations can provide detailed information about forest characteristics. Measurement techniques range in accuracy from estimations of canopy cover and tree height using the naked eye or the Cajanus tube,[4] to the use of regression models based on specific parameters.[5] However, ground-level work is time consuming and relatively expensive for a national-level inventory.[6] Remote sensing using satellites is less expensive as it covers large areas and produces rapid, consistent and transparent data (see Table 10.1).[7] Satellites provide images at three broad resolutions and these determine the minimum mapping area, cost and time to process (see Table 10.1). A combination of satellite imagery and ground observations can therefore provide a powerful tool for mapping. Box 10.1 outlines some of the techniques available for mapping and monitoring, while Box 10.2 details several case studies that have used these techniques.

Table 10.1: Satellite resolution, minimum mapping areas and costs

Resolution	Example satellites	Minimum mapping area	Cost per 1000 ha
Less than 5m	Quickbird, IKONOS	Less than 0.0025 ha	$140 to $210
20m to 30m	Landsat, ERS-2	Around 0.05 ha	$0.2 to $1.4*
250m to 1km	MODIS, MERIS	Around 20 ha	Free

* Archive data from select past years is available for free

---

4   Rautiainen et al (2005). A simple hand-held periscope like device used to look upwards with the aid of mirrors.
5   Korhonen (2006)
6   Korhonen (2006)
7   Based on Hardcastle et al (2008) estimate of $1000 and 19 days for an extensive ground survey for 1000ha.
8   Hardcastle et al (2008)

Box 10.1: Techniques for forest inventory, mapping and monitoring

A range of techniques are available for gathering and interpreting forest inventory data and monitoring land use change over time.

Remote sensing technology

Remote sensing involves obtaining information without direct contact, for example using aerial photography or satellites. Forest mapping and monitoring is principally done by satellites. There are three main types that are useful for earth observation:

- Multi-spectral: This measures reflected energy from the sun, mainly within visible and infrared wavelengths. Satellites include MODIS, Landsat and IKONOS. Using a combination of different wavelengths enables detection of many forest characteristics, and data from different satellites can then be combined. There are many satellites available at different resolutions, from one metre to 1000 metres, with daily to monthly frequencies.[8] The main disadvantage of multi-spectral satellites is that their imagery is obscured by cloud cover.
- RADAR (Radio Detection and Ranging): This emits and collects microwave and radiowaves to derive forest characteristics. There are currently relatively few RADAR satellites available. Examples include the Earth Remote Sensing Satellites (ERS 1 and 2), the Canadian Space Agency's RADARSAT and Japan's Earth Resources Satellites (JERS 1 and 2). Further projects are in development, such as the European Space Agency's Sentinel-1, due to be launched in 2011.[9] RADAR also offers some potential for measuring tree height, which can be used to calculate biomass and then carbon,[10] although this method requires further development.
- LIDAR (Light Detection and Ranging): This emits a laser pulse and calculates distance from the time elapsed to receive the reflection. This information on forest volume and structure can be used to calculate above-ground carbon estimation using mathematical relations. Currently LIDAR is available only through attachment to aircraft, so it only provides local coverage.

Remote sensing data interpretation techniques

Data interpretation techniques range from simple visual interpretation to applying complex algorithms to satellite data – and the level of training required varies accordingly. An increase in the sophistication of the technology generally increases the infrastructural costs but can reduce the processing time. However, rapid mapping can be achieved using unsupervised techniques on satellite data (that is, using an existing algorithm for classifying vegetation type without applying knowledge of the area to supervise the process), although this means there is no quality control. Supervised classification of satellite imagery can return more accurate results but requires a training phase where algorithms are adjusted and developed for the area being mapped.

On the ground measurements

Ground work is required for carbon measurements and to verify desktop forest mapping

---

8 Hardcastle et al (2008)
9 Attema (2005)
10 Neeff et al (2005)

(ie from satellite images). Field methods can simply involve collating qualitative data or gathering repeated quantitative measurements from permanent plots. Qualitative data can be quick and basic, such as noting the presence of a dominant species and relying on existing research or expert knowledge to classify or quantify data for the forest. This can be useful for verifying desktop forest mapping but produces low accuracy data for carbon measurements. Establishing permanent plots and making repeated measurements can provide accurate measurements for those plots but this method is more time-consuming and costly.

Box 10.2: Case studies of rapid forest inventory and mapping

A study of the benefits and limitations of different techniques in mapping and groundwork was commissioned for this Review from the Royal Botanic Gardens, Kew.[11]

The study describes a range of survey techniques and methodologies and provides examples of their application in a variety of contexts. These range from rapid, desktop studies using remote sensing imagery to measure trends in forest cover to full, multi-phase, desktop and ground survey methods using data on biodiversity combined with information on biomass and carbon sequestration, ecosystem services, vegetation dynamics and other features.

The Kew report includes examples of the application of the survey methods to monitoring deforestation trends and examples of using monitoring techniques in measuring past and present trends in vegetation cover and floristic content. It also discusses the applicability of these methods to predictive mapping of vegetation responses to climate change and other threats.

Example 1: Rapid mapping from satellite imagery/aerial photographs, Mount Oku, Cameroon

Rapid mapping applies a combination of expert knowledge and dated satellite imagery of the area. The approach is often used to study areas that have highly distinguishable vegetation types or where the vegetation classification is straightforward, eg forest or grassland. The major drawback with this method is that, in the absence of ground-level verification, the surveyor cannot be sure that their interpretation of the imagery truly reflects the reality on the ground. However, this is a quick and cheap way to map vegetation.

Mapping area 800Km^2
Mapping scale 1:50,000
Overview: Mount Oku and the Ijim Ridge form the largest remaining patch of montane forest in West Africa. It has exceptional levels of flora and fauna endemism – especially among birds. Since 1987 an important conservation project managed by BirdLife International has been working in the area, with the aim of reducing forest loss and improving agricultural practices. This mapping project has assessed on the ground effects of the conservation effort in this region. These were measured using traditional

---

11 Moat et al (2008)

classification of remote sensing data combined with aerial photographs for the older dates. Because of the rapid change in terrain and the varying nature of the imagery, all the data had to be corrected using elevation models derived from RADAR satellite data. The results show strong spatial patterns of deforestation between 1958 and 1988 (more than 50 per cent of the montane forest was lost in this period) followed by a regeneration period starting in 1988, just after the Conservation Project was created. In the 1988-2001 period, 7.8 per cent of the 1988 extent of montane forest had been recovered, mainly on the eastern side of the mountain.
http://www.kew.org/gis/projects/oku_cameroon/index.html

Example 2: Use of full survey techniques, Montserrat

Full surveys combine the use of satellite imagery or aerial photography with extensive ground survey work, including the collection of quantitative data on plant biomass, species density, dominance and frequency. These surveys are scientifically robust and have quality control built into them. They are also repeatable, allowing them to be used as the basis for detailed monitoring. However, they can be more expensive and time consuming.

Approximate area mapped: 100Km²
Mapping scale: 1:25,000
Field methods used: Quantitative, ground point data onto field data forms, permanent plots
Desktop methods: Statistical methods, based on elevation and climate

Overview: The Caribbean island of Montserrat was devastated by a huge volcanic eruption in 1997. This vegetation survey and map has been completed as part of the Darwin Initiative project 'Enabling the People of Montserrat to Conserve the Centre Hills'. As a result of extensive fieldwork undertaken during this project, planners and conservationists now have an accurate vegetation map and tool to help guide the island's recovery. Species and habitat monitoring continues to identify the most important areas for plant diversity. With wider collecting, the aim is to produce a conservation checklist and Red List for the whole island. Permanent biodiversity assessment points were established by the project and on-going monitoring for several threatened species is helping to further GIS and general knowledge of the biodiversity of the island at a national level.

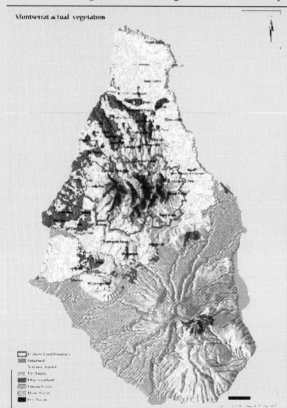

http://www.kew.org/scihort/directory/projects/EnabMontserrCentreHi.html

Example 3: Retrospective monitoring: understanding changes and trends, Madagascar

The Madagascar Vegetation Mapping project is the most notable example of Kew's involvement in large-scale monitoring of vegetation change. This project compared satellite imagery from the 1960s with imagery from the present day in order to map changes in vegetation cover. It has also established standards for the continual monitoring of vegetation into the future, with the creation of standard field data forms and panoramic photographs.

> *Approximate area mapped: 590,000Km²*
> *Mapping scale: 1:500,000 (final mapping scale) – 1:125,000 (working scale)*
> *Survey level:* Full survey, with several iterations
> *Field methods:* Quantitative, ground point data onto field data forms
> *Desktop methods:* Classification of imagery and some statistical analysis
> *Overview:* The Madagascar Vegetation Mapping Project was a four year project (2003-2007), funded by the Critical Ecosystem Partnership Fund (CEPF) and managed jointly by the Royal Botanic Gardens, Kew, the Missouri Botanical Garden, and Conservation International's Center for Applied Biodiversity Science. The project was innovative in a number of ways. It employed state of the art remote sensing technology and methodologies to delimit Madagascar's vegetation. In addition to providing the most thoroughly ground truthed vegetation map ever compiled for Madagascar – a vital baseline for future monitoring of vegetation and landscape change – the study included an analysis of past and present rates of vegetation loss for each type since the 1970s using Landsat imagery. This has helped to identify conservation priorities in a country where only about 18 per cent of primary vegetation remains.
>
> http://www.kew.org/press/madagascar_atlas.htm

Sampling reduces the costs of measuring by reducing the amount of data that needs to be collated and processed from ground work and from satellites. In particular, stratified sampling by forest type and by risk factors (such as proximity to roads, population centres and protected areas, and the suitability of the forest for other uses) can provide cost-effective results and relatively high levels of accuracy, even with low sampling intensities. The Noel Kempff Mercado Climate Action Project in Bolivia, for example, has produced carbon stock assessments with less than 5 per cent errors with only 3 per cent sampling intensity.[12]

Carbon is measured directly through lab work and using mathematical tools such as allometric equations, which relate size and shape parameters such as tree diameter to carbon stocks. Once these relationships have been established it is possible to create models or use remote sensing data (eg LIDAR) to reduce the number of carbon measurements needed. Many relationships and models exist but to produce the highest accuracy, ground work is needed to calibrate for the specific forest type.

10.3 Monitoring and verifying emissions and sequestration

Deforestation, afforestation and reforestation can be monitored effectively, using satellite technology to assess changes in forest cover and land use. Monitoring forest degradation is more challenging, although recent developments in remote sensing, coupled with ground work and additional data such as proximity to roads, can be effective.

Once the forest carbon in an area has been mapped, the next step in calculating forest carbon emissions is to monitor any changes in the area. These may be major land use

---

12 Brown et al (2000)

changes, including deforestation, afforestation and reforestation, or smaller scale changes in forest structure, including degradation and restoration. These have different monitoring requirements and are discussed below.

10.3.1 Monitoring land use change

The level of emissions released as a result of deforestation will depend on the specific land use change that has taken place, and this needs to be monitored in addition to the overall reduction in forest size. This is also the case for sequestration of CO_2 through afforestation and reforestation. Consequently, net emissions from an area are determined by the difference between the carbon stock in the original forest and the final carbon stock in the new land use that has replaced it.

A study commissioned for this Review[13] assessed the potential emissions that would result from the conversion of tropical forests to several different land use types. Figure 10.2 shows that conversion to soybean, maize or rice potentially produces 60 per cent more emissions than conversion to palm oil.

Figure 10.2: Example carbon densities of different land uses

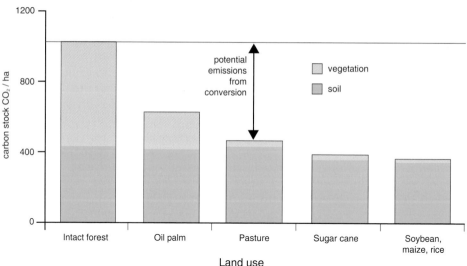

Note: Original carbon density is based on average carbon densities for tropical forests from the IPCC TAR. In reality, these vary geographically (see Figure 10.1)
Source: Miles et al (2008)

In general, land use change affects the carbon stocks held in vegetation more than those in soils. However, when the soil has a high organic content, eg in peat swamp forests, carbon emissions from the soil can be substantial following a change in land use.[14]

Emissions are released at different times and intervals according to different land uses, so sampling of time-averaged carbon densities of different land uses can improve the accuracy of emissions calculations. Plantations, for example, are regularly harvested, so carbon densities change during the cropping cycle. The nature of the end product derived from the harvested wood is also important: some wood products, such as furni-

---

13 Miles et al (2008)
14 Hooijer et al (2006)

ture, may not release the carbon held in them for over 100 years, while fuelwood may release it within hours.

Different land uses, like forest types, can be identified by satellite data, which can be used to detect properties such as leaf area and plant height of pasture and palm oil. This data is then calibrated with ground level research. Land use changes can be detected by geographically referencing data and comparing it with earlier maps. This information can then be combined with time-averaged carbon measurement samples to estimate the resulting emissions or sequestration.

The accuracy of monitoring land use changes depends on the minimum mapping area, the interpretation techniques used and the frequency of data collection. A relatively high level of accuracy is achievable. For example, a study over the Congo Basin using 30-metre multi-spectral satellite data to identify areas of deforestation and ARR between 1990 and 2000 produced 91 per cent accuracy.[15] Repeated monitoring is needed to ensure all forest changes are accounted for and attributable to a particular time period. Both Brazil and India undertake regular forest surveys. India undertakes a biennial forest survey classifying three different densities of forest canopy, scrubland and non-forest areas at the one hectare scale over the entire country.[16] Verification using higher resolution satellite imagery found the accuracy of surveys to be 92 per cent.[17] Brazil has a system not only for monitoring annual changes but also for lower resolution bimonthly assessments to identify and prevent illegal deforestation (see Box 10.3). This demonstrates the feasibility of monitoring land use change at a national level.

Box 10.3: Detecting deforestation in the Brazilian Amazon

The Review visited Brazil's National Institute for Space Research (INPE) which is tasked with assessing the extent and rate of deforestation in the Brazilian Amazon, an area covering approximately 5 million square kilometres, using satellite images. INPE produces this data in partnership with the Ministries of Science and Technology, Ministry of Environment and IBAMA, the environmental enforcement arm of the Brazilian government.

DETER (Near Real Time Deforestation Detection)

DETER allows estimates to be made of deforestation based on data gathered every few days.[18] It uses high frequency observation satellites in order to reduce complications due to cloud cover. The high frequency of data allows DETER to supply IBAMA with bi-monthly information on deforestation. Data is published monthly on the website www.obt.inpe.br/deter. However, the resolution of DETER's sensors is not sufficiently high to estimate the total area of cleared land. For that, INPE employs better resolution images produced by PRODES.

---

15 Brown et al (2008)
16 Brown et al (2008)
17 Brown et al (2008)
18 MODIS satellite data is used. NASA (2008).

> **PRODES (Annual Deforestation Rate Assessment)**
> PRODES calculates the yearly consolidated deforestation rate in the Brazilian Amazon. The satellites work on a 16 day and 26 day re-visit and provide precise images (within a range of 20-30m),[19] making it possible to detect any deforested area larger than six hectares.
>
> **DETEX (Forest Exploitation Detection System)**
> This programme was established in 2007 and allows rapid intervention by members of the federal environmental agency. The system collects images of 20 x 20m areas - in contrast to PRODES (30x30m) and DETER (250x250m) – with the purpose of detecting clearings in the forest due to selective logging activities.
>
> INPE has a stated policy of free public access to all its deforestation data.
> Brazil is one of only a few forest nations with the capacity to analyse satellite data.[20] However, there is potential for Brazil to share this expertise with other tropical forest countries, to support the development of their own forest monitoring capacity.

10.3.2 Monitoring forest degradation

Changes in forest carbon caused by forest degradation can be more difficult to detect than land use changes. Information about the type as well as the extent of degradation is required in order to estimate emissions. Intensive ground work can be used to detect forest degradation, although this is time consuming and costly over large areas. Degradation that creates a gap in the forest canopy, such as selective logging, can be detected using very high resolution remote sensing, although this, too, can be costly (see Table 10.1). However, significant progress has been made in the last ten years in detecting forest degradation using combined satellite and ground work approaches. For example:

- The Carnegie Landsat Analysis System uses mid-resolution, multi-spectral Landsat satellite data and atmospheric modelling to detect land use change in the Amazon. It can detect selectively logged areas in the Amazon with 86 per cent accuracy.[21]
- Degradation can be inferred by proximity to roads, which can be detected by mid-resolution satellite data. A study in southern Cameroon found 80 per cent of anthropogenic forest disturbance was within two kilometres of roads.[22]
- Souza et al have developed the Normalized Difference Fraction Index, derived from mid-resolution multi-spectral satellite data, to identify degraded forest, intact forest, forest regeneration and recently and past logged areas. Their study of the east Amazon using this technique produced a 93 per cent accuracy[23] level when compared with the very high resolution IKONOS data.

---

19 Landsat data is primarily used which has a repeat pass of 16 days. This is complemented by the use of other satellite data such as CBERS with a 26 days repeat pass.
20 Hardcastle et al (2008)
21 Asner et al (2005)
22 Mertens and Lambin (1999)
23 Souza et al (2003)

These examples illustrate that the accuracy of techniques to detect forest degradation has improved considerably. However, further improvements could be made and more research and development should be undertaken. Signs of forest degradation become harder to detect from satellite data after one or two years, and so satellite data should be collated regularly to ensure robust carbon accounting.[24]

Some types of degradation do not create gaps in the canopy, for example gathering of deadwood and under-storey vegetation. This makes these activities much more difficult to detect, although factors such as proximity to population centres can be used as an indicator. While these types of degradation tend to result in lower-impact, short-term changes,[25] further research should be undertaken in this area and in the area of monitoring forest restoration, which can use similar techniques to those for monitoring degradation.

10.3.3 Monitoring and policy development

Monitoring data can be used not only for detecting land use change and forest degradation, but also for policy development. Understanding where deforestation or degradation rates are highest and correlating that information with socio-economic data may help policymakers identify the underlying drivers of deforestation and degradation.

10.4 International and national approaches to measuring and monitoring

The IPCC has developed internationally agreed measures and guidance and these should be used to achieve consistency in monitoring and transparency in reporting.

Significant progress has been made on international standards for national-level forest inventories, and the IPCC Good Practice Guidance has resulted in major improvements in developed country estimates of agriculture, forestry and other land use. However, capacity building and national-level research will be needed in many other countries.

In countries where capacity is still developing, methodologies used may be less accurate. In these cases, conservative estimates of emissions and emissions reductions should be used.

International standards and methodologies for measuring and monitoring are needed in order for all countries to be able to produce comparable national emissions inventories. These should enable countries with differing levels of expertise and capacity to calculate emissions that are as comparable as possible. The IPCC Good Practice Guidance has resulted in major improvements in developed country estimates of agriculture, forestry and other land use (AFOLU – previously known as LULUCF). However, capacity building and national-level research will still be needed in many countries.

---

24 Souza et al (2003)
25 The gathering of deadwood for fuel, for example, has relatively low impact because deadwood generally represents less than 10 per cent of forest carbon stocks, Brown et al (2008)

10.4.1 National emissions inventories under the UNFCCC

Under the Kyoto Protocol, developed countries are required to use internationally agreed guidance provided by the IPCC for estimating emissions from land-use, land use change and forestry, and to submit this data on an annual basis (see Chapter 7). Developing countries are also encouraged to use the IPCC methods for the inventories that they produce on a periodic basis.

The introduction of Good Practice Guidance (GPG) by the IPCC in 2003 has resulted in major improvements in developed-country AFOLU estimates (see Box 10.4)[26] and use of the IPCC guidance is increasingly widespread. As a means of estimating carbon stocks and emissions, it also forms a useful basis in international agreements which seek to extend incentives to reduce emissions from deforestation and forest degradation in developing countries, and to encourage conservation and sustainable forest management.

Box 10.4: Good Practice Guidance for national GHG emissions inventories

The IPCC's Good Practice Guidance for Agriculture, Forestry and Other Land Use includes specific guidelines for the forest sector.

The guidance assists countries in producing inventories for AFOLU that are neither overestimates nor underestimates so far as can be judged, and in which uncertainties are reduced as far as practicable. It supports the development of inventories that are transparent, documented, consistent over time, complete, comparable, assessed for uncertainties, subject to quality control and quality assurance, and efficient in the use of resources.

The methods set out in the GPG are graduated into three tiers to accommodate national circumstances. The most basic (Tier 1) contains default methods. Tier 2 has the same mathematical structure as Tier 1 but uses more country specific data, while Tier 3 provides flexibility for more sophisticated methods. The tiers enable estimates to be calculated with different levels of resources, while maintaining consistency.[27] For example, in calculating emissions from the conversion of forest to pasture, all forest carbon stores may produce emissions but changes in above-ground carbon produce the greatest emissions. So Tier 2 or Tier 3 should be applied, while litter, deadwood and soil carbon produce smaller total emissions and Tier 1 could be used if resources are limited.

The guidance estimates deforestation as the sum of transitions from forest to other land uses. Degradation would be estimated as long-run decline in carbon stocks, using the methods for estimating emissions, removals and carbon stock changes associated with forest management. GPG methods are also suitable for estimating total forest carbon stocks and changes associated with sustainable management of forests. The guidance covers all relevant biomass and dead organic matter carbon pools, including litter and soils. There are methods for estimating uncertainties in annual estimates and trends that could be used for conservative accounting which some have advocated as a way to incentivise use of more accurate (in general, higher tier) methods.

---

26 In 2006 the IPCC produced the 2006 Guidelines for Agriculture, Forestry and Other Land Use (AFOLU). The Parties are currently discussing the adoption of the 2006 Guidelines. The Review therefore refers to the GPG which is the currently agreed guidance. The 2006 Guidelines use the GPG concept and the methods are consistent. In some cases the 2006 Guidelines contain updated data and can be used as a scientific source when applying the GPG

27 IPCC (2006)

Many developing countries have less experience of applying IPCC guidance than developed countries. Some forest nations currently have little or no data for mapping forest carbon stocks, and capacity building and national level research will be required (see Section 10.5).

In 2005, the UNFCCC initiated a REDD process that included providing further clarification on the IPCC guidance for reporting forest carbon emissions and sequestration at the national level. GOFC-GOLD is the ad hoc REDD working group that focuses on providing international consensus on methodological issues relating to quantifying carbon impacts of implementation activities specifically for REDD. The working group has produced a first draft of a methodological sourcebook.[28]

In countries where capacity is still developing and methodologies used may be less accurate, conservative estimates of emissions and emissions reductions should be used to increase the confidence that targets have been complied with. In this case, for example, accounting of emissions from deforestation would be at, say, an upper agreed percentile of the uncertainty range rather than at the central estimate. This would provide an incentive to introduce more accurate estimates and reduce the cost and capacity barriers for countries to participate in forest mitigation activities.

10.4.2 Sub-national programmes

Some of the poorest countries will not be able to build sufficient capacity in the short term. For these countries, similar methods described for national-level measuring and monitoring could be applied to sub-national programmes.

Measuring and monitoring forest emissions and sequestration at project and programme level is similar to that undertaken at national level. The IPCC GPG also provides advice for use at this level. The uncertainties associated with intra-national leakage would be expected to be greater when measuring and monitoring at the sub-national level and this may have implications for the types of incentives that should be made available (see Chapter 9).

10.4.3 Review, verification and accreditation

Review and verification of forest emissions are important to ensure that reductions are additional and long-lasting. Satellite images of changing forest cover provide a greater degree of transparency in monitoring forest emissions reductions than monitoring in any other sector.

Review and verification of forest emissions are important to ensure that reductions are additional and long-lasting. Currently, procedures for national emissions inventories exist under the UNFCCC for Annex I countries. These are reviewed annually by a group of UNFCCC experts, with in-depth country visits every three or four years. This ensures consistency and transparent use of GPG methods. International negotiations will need to determine the level of review and verification required for reductions in forest emissions in developing countries as is done for developed countries. However, any system of national monitoring, reporting and verification will need to follow some broad principles.[29] International buyers will need assurance that real emissions reductions have

---

28 GOFC-GOLD (2008)
29 Peskett and Harkin (2007)

actually occurred, through rigorous monitoring, periodic review and evaluation of institutional performance. To that end, the use of satellite data to monitor forest emissions potentially provides a high degree of transparency that is not possible in other emitting sectors.

In addition to monitoring, reporting and verification, accreditation will also require an institutional process. Various functions will need to be performed. Emissions reductions and their sources will need to be recorded in the national registry. Emissions reductions will then need to be converted into tradeable credits, and the movement of credits tracked through an international transaction log (see Chapters 7 and 12 for more details on the current international system and the distribution of credits and finance at a sub-national level).

10.5 Capacity building: expertise and costs

An estimated $50 million will be needed for a sample of 25 forest nations to set up national forest inventories, with a further $7-17 million needed for annual running costs. Technological advances, knowledge sharing and international cooperation can reduce these running costs over time.

The capacity of forest nations to measure and monitor their forest emissions varies greatly. Many have little or no up to date data for mapping carbon stocks. Forty per cent of countries' most recent assessment of forest area was over ten years ago and a third have no national-level data available for carbon stock estimation.[30] Capacity building and national-level research are required to enable all forest nations to estimate emissions and sequestration accurately and cost-effectively. In particular, the application of IPCC methods should be a priority and this will involve substantial capacity building efforts.

Capacity building at national level could significantly improve the accuracy of measuring and monitoring. Using IPCC Tier 1 global biome averages with no national level research can produce carbon stock accuracies of around 50 per cent[31] (see Box 10.4 on IPCC Tiers). With appropriate techniques, sampling of 3 per cent of the forest area can significantly improve accuracy.[32] Evidence suggests that this level of sampling can give carbon stock accuracies of up to 95 per cent provided good ground based data are available on ecosystem carbon densities (see previous sections). Using conservative estimates where levels of uncertainty are high will provide a further incentive to develop more accurate national forest emissions inventories.

Research was commissioned for this Review to estimate the costs of setting up and running national forest emissions inventories (including establishing a baseline) based on the techniques and IPCC guidelines described in the previous sections.[33]

The findings of the research are based on a hypothetical scenario. This was a medium-sized country with 50 million hectares of forest, with no existing forest inventory or remote sensing capacity and where no data had previously been collected. A sample of 25 forest nations (all developing countries) was then chosen to illustrate costs for different circumstances, and compared with the reference scenario. The 25 countries

---

30 Based on FAO FRA (2005)
31 Brown et al (2008)
32 Brown et al (2008)
33 Hardcastle et al (2008)

chosen together account for nearly 40 per cent of the world's forest cover. The existing capacity and data for these 25 countries were assessed and compared with the costs for the reference scenario in order to estimate individual costs for each country. The costs of setting up and running an emissions inventory in this reference country were estimated to be in the order of $2 million in the first year and $0.7 million per year (upper estimate) thereafter.[34] This gives a figure of $50 million for the 25 forest nations to set up national forest emissions inventories, with an additional $7-17 million needed to cover annual running costs (the higher costs include forest degradation).[35] These figures are incorporated into the overall estimate for policy and institutional reform costs for 40 countries detailed in Chapter 13.

Several international organisations are already providing and planning support to developing countries to increase their measuring and monitoring capability. Since 2000, the FAO National Forest Monitoring and Assessment Programme has been helping countries design and implement the collection of forest and land use information. Assessments have been set up in 18 countries, and are planned in at least a further 11.[36] The assessments gather information on: the extent of forest types and land use, growing stocks, environmental problems, biomass and carbon, biological diversity, use and management of the forest resources and non-wood forest products. In some countries the assessment is extended to an Integrated Land Use Assessment (ILUA) by collecting information on other land uses such as crops, livestock, soil and water.[37] Capacity building to enable countries to update, expand and manage their forestry information base is a main focus of the programme.

The World Bank Forest Carbon Partnership Facility (FCPF) Readiness Fund has identified 14 countries for initial support, specifically to help them prepare for participation in an international forestry mechanism under the UNFCCC. The facility will provide financial and technical support to countries to develop baselines, set up a monitoring system and develop a strategy for tackling deforestation at a national level. The FCPF is discussed further in Chapter 13.

The capacity building costs of setting up robust national forest emissions inventories are likely to be relatively low. Furthermore, costs could be reduced further through international cooperation and technological advances. For example, forest monitoring systems can have a high cost per hectare in countries with small forest areas as there is a minimum level of technical requirements and associated costs. Small countries could partner with other countries to share this cost. For example, in Central Africa there are several countries with small forest areas. Work commissioned by this Review has estimated that a regional monitoring partnership between Cameroon, DRC, Congo, Equatorial Guinea and Gabon in Central Africa could save more than $2.2 million in setup costs in the first year, and more than $0.5 million in annual running costs.[38] Congo Basin countries are already cooperating on forest monitoring as part of COMIFAC.[39]

---

34 At £:$ exchange rate of 1.84, September 2008
35 Hardcastle et al (2008)
36 Saket (2008)
37 FAO (2008)
38 Hardcastle et al (2008)
39 COMIFAC: Commission des Forets d'Afrique Centrale

10.6 Conclusion

Robust measuring and monitoring is essential to ensure that credits are awarded for genuine reductions in national emissions. The IPCC Good Practice Guidance should be used to achieve consistency in monitoring and transparency in reporting. Evidence shows that if appropriate techniques are used and the guidance is followed, measurements of forest emissions can reach levels of accuracy of 90 per cent,[40] consistent with levels achieved in other inventory sectors.[41] Forest emissions can be measured and monitored with higher levels of confidence than emissions from sources such as fugitive emissions from fuels and nitrous oxide.[42] Satellite images of changing forest cover also provide a greater degree of transparency in monitoring forest emissions reductions than monitoring in any other sector.

In countries where measuring capacity is still developing, conservative estimates of emissions and emissions reductions should be used to ensure that credits are awarded for real reductions achieved.

Despite significant progress on international standards in measuring and monitoring emissions, capacity building will be necessary. However, the costs of this capacity building are relatively low. Further advances in satellite and computer technology will be important along with training, international cooperation and knowledge sharing.

The next chapter sets out how the emissions reductions monitored can be converted to credits and linked to emissions trading schemes.

---

40 Using IPCC GPG Tier 1 uncertainity analysis and values of 95 per cent and 91 per cent accuracy respectively for measuring carbon stocks and monitoring, produces an overall certainty of 90 per cent. See Brown et al (2008) for uncertainty analysis and later sections in this chapter for evidence of accuracy levels achievable.
41 The EC (2004) found fossil fuel combustion emissions traded on the EU ETS to have over 90 per cent certainty estimates.
42 Monni et al (2007)

11. Linking to carbon markets

Key messages

The third building block for tackling forest emissions in the medium term is a well-designed mechanism for linking forest abatement to carbon markets, and accessing funding from the private and public sectors as carbon market finance grows.

Linking forest abatement to carbon markets has potential implications for the:

- price incentive to invest in new low carbon technologies;
- transfer of existing cleaner technologies to developing countries;
- level of finance required in the medium term from non market sources for significant forest sector abatement.

There is a large potential supply of forest sector abatement in developing countries: modelling suggests 3.5 $GtCO_2$ per year in 2030. This potential abatement needs to be tapped to give the world a realistic chance of limiting global temperature rise to 2°C.

Demand for international credits is likely to grow as Annex I countries commit to more stringent emissions reductions targets and new trading schemes are created.

This Review modelled impacts on (a) the EU carbon market price and (b) the international credit market using three variables:

- supplementarity limits (the proportion of credits from non-Annex I countries that Annex I countries and companies are permitted to use to meet their targets);
- emissions targets;
- whether or not forest credits are admitted to the international credit market.

Impact on the EU ETS carbon market price

Raising a carbon market's supplementarity limit lowers its carbon price. Although this improves efficiency in the short term, it could potentially reduce the price incentive to invest in new low carbon technology for the long term. However, if Annex I countries simultaneously commit to more stringent emissions targets, then the carbon price can be maintained or even rise.

Modelling results suggest that the EU carbon market price would be similar during Phase III whether (a) Member States committed to a 20 per cent emissions cut with a 30 per cent supplementarity limit or (b) committed to a 30 per cent emissions cut with a 50 per cent supplementarity limit.

By adopting both more stringent emissions targets and a higher supplementarity limit, the EU could achieve a greater overall reduction in its emissions and drive more abatement in developing countries without increasing the costs to European industry. Getting the balance right is key.

Modelling suggests that if supplementarity limits are set at 50 per cent or lower in Phase III of the EU ETS, then admitting forest credits into the international credit market would have little or no impact on the EU carbon market price. This is because, when restrictions on the use of non-Annex I country credits are this tight, more costly EU abatement would still be necessary and would continue to set the price for all units of abatement in the carbon market.

Impact on the international credit market
If the forest sector is able to realise a significant amount of its abatement potential, then modelling suggests that forest credits could constitute at most 34 per cent of the international credit market in 2020.

The international community needs to balance emissions targets and supplementarity limits in order to achieve the desired trade-off between four objectives:

- fund significant forest abatement;
- reduce the cost to industry of meeting more stringent global emissions targets;
- provide a strong incentive to invest in new clean energy technologies;
- support a high level of technology transfer to the developing world.

Impact on the level of non-carbon market finance required to reduce forest emissions
The choices that the international community make on the stringency of global emissions targets, supplementarity limits and credit design features will have a large impact on the amount of carbon market finance available for reducing forest emissions.

One scenario modelled suggests that the global carbon market could supply around $7 billion per year for forest abatement in 2020. This would leave a funding gap of around $11-19 billion for halving forest emissions, which would need to come from other private and public sources.

Linking mechanism
During the transition to a comprehensive global cap and trade system, a linking mechanism could perform three important functions. First, a single institution should be used to aggregate different sources of funding and purchase all the forest credits generated by forest nations. Second, a proportion of the credits generated should be placed within a reserve and used to offset any future reversal of forest emissions reductions. Third, the institution could also be used to reduce the risk of investing in forest abatement for forest nations by guaranteeing to purchase from them a minimum quantity of credits at a minimum price.

11.1 Introduction

The third building block for tackling forest emissions in the medium term is a well-designed mechanism for linking forest abatement to carbon markets, and accessing funding from the private and public sectors as carbon market finance grows

Linking forest abatement to carbon markets has potential implications for the:

- price incentive to invest in new low carbon technologies;
- transfer of existing cleaner technologies to developing countries;
- level of finance required in the medium term from non-market sources for significant forest sector abatement.

Under a comprehensive global cap and trade system, the system of international supply and demand for forest abatement would be self-regulating. Countries that delivered more abatement from the forest sector than they required to meet their own emissions cap would be able to sell on the excess to countries (or companies within those countries) that were unable to meet their own emissions cap domestically. However, in the medium term a transitional system that combines finance from carbon markets and other sources (set out in Chapter 8) is likely to depend on, and have important implications for, the supply and demand of international credits, including those from the forest sector.

A potentially large supply of forest credits could be available to enter the global credit market relative to current demand from Annex I countries. Demand is set to increase markedly, but will depend on the stringency of emissions reduction targets that Annex I countries commit to in a global climate change deal as well as the limits they set on how far those targets can be met by the purchase of non-Annex I credits (supplementarity limits). This chapter explores the potential impact of these two variables, and of linking forest abatement to carbon markets, on the carbon price of the EU Emissions Trading Scheme (EU ETS) and on the international credit market. When a supplementarity limit is described as being low in this Review, we mean that the restriction on international credits that can enter a carbon market is tight (and visa versa).

Stringent targets are necessary to give the world a realistic chance of limiting global temperature rise to 2°C. Linking forestry abatement to carbon markets is efficient because it reduces the cost of global climate change abatement and should enable the world to commit to more stringent emissions reduction targets (see Chapter 6). However, impacts on Annex I carbon market prices have implications for the incentive to develop new low carbon technology, which will be essential for efficient global climate change mitigation in the long-term. And certain sectors of the international credit market contribute to the transfer to developing countries of existing low carbon technologies which help avoid lock-in to high carbon development paths.

This chapter sets out the results of modelling the impacts of linking forest abatement to carbon trading schemes. Based on this modelling, the chapter goes on to explore the impact of both the stringency of a global emissions target and supplementarity limits on the level of finance the carbon market could generate for forest emissions abatement over the medium term, and consequently the level of additional funding needed from other private and public sources. Finally, we examine how these funding sources might be best combined and used to manage the risk of reversal of forest emissions reductions, as well as reduce the risks to forest nations investing in emissions reduction policies.

11.2 Carbon markets: supply and demand

Carbon markets rely on a few basic tenets:

- as long as standards are applied consistently, the location of abatement is irrelevant to the global goal of stabilising GHGs (since they mix in a shared global atmosphere);
- mitigation potentials and costs vary across countries and sectors;
- trading of allowances and credits between countries or companies unlocks cheaper abatement opportunities and reduces the overall cost of reaching a given global goal;
- the greater the geographical and sectoral scope of the market, the cheaper abatement will be for any given emissions reduction target.

Within the UNFCCC trading system, encompassing three flexible mechanisms, there are various buyers and sellers (see Box 11.1). Under a scheme that included the forest sector fully in a global trading system, forest nations would become sellers of credits, adding to the supply of emissions reduction credits in the market. At the same time, increased demand in the system would need to come from Annex I countries taking on tighter emissions caps. Before examining how to link forest credits to emissions trading schemes, it is important to understand the credits that are being supplied to the market and the demand for those credits in the market. The following sections set out some key features of the supply side and the demand side.

11.2.1 Forest credits: the supply side

There is a large potential supply of forest sector abatement in developing countries: modelling suggests 3.5 GtCO$_2$ per year in 2030. This potential abatement needs to be tapped to give the world a realistic chance of limiting global temperature rise to 2°C.

As set out in Chapter 9, national level reference levels, or baselines, will need to be set for forest nations participating in an international climate change framework. Credits can be awarded when net emissions from the forest sector in the country (including REDD and ARR) fall below the national reference level. These credits can then be traded in national and regional carbon markets (see Figure 11.1).

The supply of credits from the forest sector in developing countries is potentially large. Modelling for this Review predicts that REDD could represent an annual emissions reduction of 2.6 GtCO$_2$ in 2030, while ARR could represent a further reduction of 0.9 GtCO$_2$ against business as usual.[1] As a result, the forest sector could be delivering 3.5 GtCO$_2$ of climate change abatement by 2030 (see Chapter 6).

---

1 There would not necessarily be a corresponding supply of credits, as BAU would not necessarily be used as the credit reference level. Further, the reference level would be likely to become more stringent over time (see chapter 9).

Figure 11.1: The supply of credits to the carbon market

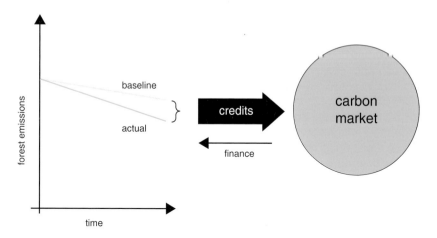

As set out in Chapter 6, the key advantages of including the forest sector in a global emissions trading system are both the level of ambition that can be achieved and the potential for increased efficiency in reducing costs of achieving reductions. This is because large emissions reductions can be achieved in the forest sector at relatively low cost compared to the more expensive abatement alternatives in other sectors (at least initially, after which abatement costs may rise more sharply – see Chapter 5). During a transition period over the medium term, it is unlikely that all forest nations will be ready to participate fully in a national forest emissions reduction scheme due to capacity constraints (see Chapter 13). This would limit the supply of forest credits into markets. Nevertheless, the remaining supply could still be considerable.

11.2.2 Emissions trading schemes: the demand side

Demand for international credits is likely to grow as Annex I countries commit to more stringent emissions reductions targets and new trading schemes are created.

Annex I countries are committed to making emissions reductions under the Kyoto Protocol[2] and have various means available to them. They can reduce domestic emissions using direct policy instruments, such as company emissions trading schemes (eg the EU ETS), carbon taxes, standards and subsidies. In addition, Annex I countries can purchase surplus national emissions allowances from other Annex I countries, or purchase emissions reduction credits from other Annex I countries (Joint Implementation) or from non-Annex I countries (Clean Development Mechanism). These mechanisms are discussed in Chapter 7.

It is demand for credits and allowances that is relevant here, as forest credits would be competing with other sources of supply in the carbon market. The projected total demand for credits and allowances over the period 2008-12 is low, at only 2.4 GtCO$_2$e. Some of the reasons for this are discussed in Box 11.1.

---

2 Except the US, which signed but did not ratify the Protocol.

Box 11.1: Demand for allowances and credits in the Kyoto market

Demand for emissions allowances and credits from Annex I countries in the first Kyoto commitment period is fairly low compared to potential supply. A number of industrialised countries received generous allowances which did not provide substantial disincentives to reduce emissions, while not all countries ratified the Kyoto Protocol. Emissions cuts will need to be much deeper post-2012 if the world is to have a realistic chance of limiting temperature rise to 2°C, and the full inclusion of the forest sector can make this more affordable.

Figure A shows projected demand for allowances and credits in the period 2008-12. In total, the demand is estimated at roughly 2.4 GtCO$_2$e. Surplus allowances (Assigned Amount Units-AAUs) from economies in transition may be as high as 7.3 GtCO$_2$e, which would be sufficient to meet the demand three times over. In practice, such an oversupply is unlikely to occur because in the EU ETS – where most compliance market trades take place – companies cannot use AAUs to meet their obligations. Other potential buyers may avoid certain AAUs that carry a reputation risk of having little or no environmental additionality. At the same time, potential supplier countries may choose to bank surplus AAUs for future, more demanding, compliance periods.

Figure A: Projected supply and demand under the UNFCCC in the period 2008-12

Potential demand from industrialised countries (2008 - 12)		Potential supplies (2008 - 12)		
Country or entity	KMs demand (MtCO$_2$e)		Potential surplus of AAUs (MtCO$_2$e)	Potential GIS (MtCO$_2$e)
EU	1,940	Russian fed	3,330	(0<???)
government (EU-15)	540	Ukraine	2,170	(1,000-1,200)
private sector (EU ETS)	1,400	EU-8+2	1,720	(100-700)
questionable P&Ms	(200)	Other EITs	85	???
Japan	450	TOTAL	7,305	(1,100-1,900)
GoJ	100			
private sector	350			
additional demand	(200)			
Rest of Europe and New Zealand	45	CDM and JI potential (MtCO$_2$e)		
government	20			
private sector (Norway and NZ ETSs)	25			
additional demand	(20)	CDM	1,600	(1,400-2,200)
Australia	0	JI	230	(180-280)
TOTAL	2,435	TOTAL	1,830	(1,580-2,480)
government	660			
private sector	1,775			
additional demand	(420)			

Source: World Bank (2008)

A number of national and regional carbon markets are proposed or have recently started to operate (see Figure 11.2 and Box 11.2). The EU ETS is currently the largest of the markets in operation and consequently is responsible for the majority of demand for credits. However, the number of additional schemes is set to grow, including schemes in New Zealand, Australia, Japan and Canada (see Box 11.2).

Although a national cap and trade scheme does not currently exist in the US, there have been a number of legislative proposals (such as the Warner-Liebermann Bill) seeking to establish one. In its absence, some states are setting up their own cap and trade schemes that will allow them to trade with each other (see Figure 11.2).

Figure 11.2. Growth of nationals and regional emissions trading schemes

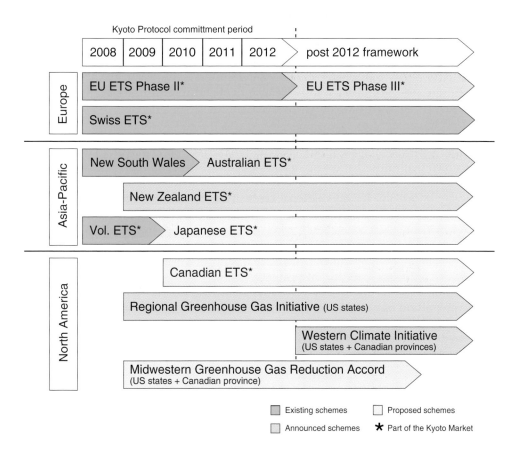

In Chapter 2 we saw that the world will need to stabilise the concentration of atmospheirc GHGs between 445 and 490ppm CO_2e in order to have a realistic chance of limiting global temperature rise to 2°C. This will require significantly larger cuts in global GHG emissions. Such cuts would create a much larger demand for forest credits in the medium term than currently exists. And from a different angle, Annex I countries are likely to agree to more stringent emissions reductions than would otherwise be the case if they know that they will be able to access a quantity of reasonably cheap credits from the forest sector.

Increasing the proportion of non-Annex I credits that Annex I countries and companies can use to meet their emissions targets (ie increasing the supplementarity limit) would also increase demand for these credits (including forest credits). See Section 11.3 for a discussion of the potential implications of increasing supplementarity limits.

The growth in demand for international credits in the markets in Figure 11.2 will be an important factor in the success of linking forest credits into emissions trading schemes. The Energy Modelling Forum (EMF) estimates that in 2020 demand for international credits could be 3.2 $GtCO_2$ per year, representing transactions of $25 billion per year.[3] The 3.2 $GtCO_2$ per year figure is the median value from a large range and the modelling assumes Annex I countries reduce their GHG emissions by 20 per cent relative to 1990 levels.

By 2030, the EMF estimates that demand for international credits could have grown to 6.4 $GtCO_2$ per year, representing transactions of $107 billion per year. In this scenario, it is assumed that Annex I countries reduce their GHG emissions by 30 per cent relative to 1990 levels.

However as demand for international credits from non-Annex I regions will depend on the stringency of the global emissions target, any supplementarity restrictions that Annex I markets place on carbon credits and the cost of abatement in non-Annex I regions, demand estimates vary. The core scenario modelled in sections 11.3.3 and 11.4 estimates demand for non-Annex I credits at around 1.7 $GtCO_2$ in 2020 and 3 $GtCO_2$ in 2030, which are lower than the EMF estimates.

In Section 11.4 we discuss the results of modelling commissioned by this Review on the proportion of potential forest abatement that the carbon markets could finance in the medium term. But before then, we examine the potential impact of introducing forest credits on the EU carbon market price as well as the impact on the international credit market.

---

3   These estimates are reported in UNFCCC (2007)

Box 11.2: Current and proposed emissions trading schemes (2008)

European Union
The EU Emissions Trading Scheme (EU ETS) was the first international ETS for greenhouse gas emissions in the world.[4] The cornerstone of EU climate change policy, it also currently accounts for the majority of the global carbon market. It covers over 11,500 energy-intensive installations across the EU, which represent close to half of Europe's emissions of CO_2. Member States transfer part of the effort required to meet their Kyoto commitments onto the private sector entities directly responsible for those emissions. The installations may choose to reduce emissions internally or, purchase EU allowances or international Kyoto credits. Norway, Iceland and Liechtenstein joined in 2008.

New Zealand
New Zealand has launched a national ETS to cover all sectors in the economy and the six Kyoto GHGs.[5] It will constitute New Zealand's core price-based measure for reducing GHG emissions and enhancing forest carbon sinks. The ETS will allow both sales to, and purchases from international credit markets, with no planned supplementarity limit in addition to Kyoto constraints. This will reduce abatement costs, aid liquidity in its relatively small market and act as a safety valve on price. The forest sector was the first sector to enter the ETS; its first compliance period is from January 2008 until December 2009. This involves the devolution to landowners of both the credits for forest activities that lead to a removal of CO_2 from the atmosphere and the liabilities for subsequent release of CO_2 through deforestation.

Switzerland
Companies in Switzerland that do not wish to pay a carbon tax for energy-related emissions may opt into an emissions trading scheme after agreeing a legally binding emissions target with the government.[6] International credits may be used to cover a maximum of around 8 per cent of the target. Switzerland has indicated that its scheme could be linked to the EU ETS in the future.

Australia
Australia has announced plans for a carbon trading scheme.[7] The government proposes to include the forest sector on an 'opt in' basis from the start. As forests in Australia are estimated to currently sequester more carbon than they emit, forest landholders would have an incentive to voluntarily include their forests in the scheme. Forest landholders will be issued allowances for the net increase in CO_2 stored in forests. There would be a liability for net reductions in stored CO_2, consistent with UN accounting rules. In the short term, principally to minimise implementation risks, the government proposes that there will be limits on the number of international offset credits that firms can use.

---

4 http://ec.europa.eu/environment/climat/emission/index_en.htm
5 New Zealand Government (2007)
6 http://www.bafu.admin.ch/emissionshandel/index.html?lang=en
7 Australian Government (2008)

> **Other proposed emissions trading schemes**
> Japan is proposing a pilot ETS to begin in 2008. Currently there is a national industry-led voluntary ETS (Keidanren Voluntary Action Plan) where participants commit to a reduction in emissions (relative to a baseline) in exchange for a subsidy to install emissions reduction equipment. Sector-based pledges are not legally binding, but a company that misses its target has to pay a penalty to the government in the form of emissions offsets, approved by the UNFCCC, from projects overseas. Canada is also considering a federal emissions trading scheme.

11.3 Price impacts of linking forest credits to emissions trading schemes

This Review modelled impacts on (a) the EU carbon market price and (b) the international credit market using three variables:

- supplementarity limits (the proportion of credits from non-Annex I countries that Annex I countries and companies are permitted to use to meet their targets);
- emissions targets;
- whether or not forest credits are admitted to the international credit market.

Linking forest abatement to global carbon markets has the potential to reduce the carbon price in Annex I carbon markets. From the point of view of short-term efficiency this would be desirable as it would bring down the cost of climate change abatement and enable the world to agree more stringent emissions targets than would otherwise be the case. However, lowering the carbon price would not necessarily be efficient in the long term as it would weaken the price signal incentivising investment in new low carbon technologies. It is only by developing and commercialising effective and affordable clean energy technologies that climate change mitigation costs can be reduced in the long term and for the world to have a realistic chance of limiting global temperature rise to around 2°C.

As well as affecting the carbon price in Annex I markets, introducing forest credits to the international credit market also has the potential to displace abatement in other sectors in the developing world. For example, if the size of the power sector (currently a component of the Clean Development Mechanism) were diminished, this could reduce the amount of finance for cleaner energy technology flowing to the developing world. And because of the long operating life of power technology, this could contribute to developing countries being 'locked in' to high carbon development paths.

Although these potential impacts present a challenge for linking forest abatement to carbon markets, this section shows that they can be addressed with a well-designed system. In particular, Annex I countries must take on sufficiently stringent emissions targets post-2012, balanced against appropriate supplementarity limits.

To test the potential for such impacts, and how they can be avoided, this Review commissioned three pieces of market modelling[8] to study the:

a) impact on the EU Allowance (EUA) price[9] of different GHG emissions caps and supplementarity limits for the use of international credits (without forestry);
b) impact on the EUA price of including forestry credits within the mix of international credits, once again against various emissions caps and supplementarity limits; and
c) share of the international credits market that the forest sector could make up.

The results of these three pieces of modelling are presented and discussed in the three subsections below.

11.3.1 Price impacts of international credits on the EU ETS

Raising a carbon market's supplementarity limit lowers its carbon price. Although this improves efficiency in the short term, it could potentially reduce the price incentive to invest in new low carbon technology for the long term. However, if Annex I countries simultaneously commit to more stringent emissions targets, then the carbon price can be maintained or even rise.

Modelling results suggest that the EU carbon market price would be similar during Phase III whether (a) Member States committed to a 20 per cent emissions cut with a 30 per cent supplementarity limit or (b) committed to a 30 per cent emissions cut with a 50 per cent supplementarity limit.

By adopting both more stringent emissions targets and a higher supplementarity limit, the EU could achieve a greater overall reduction in its emissions and drive more abatement in developing countries without increasing the costs to European industry. Getting the balance right is key.

As greenhouse gases mingle in the atmosphere, it makes no difference to climate change where in the world they are emitted or reduced. Abatement in developing countries is generally cheaper than abatement in developed countries. So it could be argued that there should be no limits on supplementarity because the more international abatement that can be used by Annex I countries to meet their targets, the cheaper their mitigation costs will be, enabling them to commit to more stringent emissions targets.

However, Annex I countries often also rely on their carbon market price to achieve technology policy goals. The higher the price, the greater the incentive is to industry to invest in new low carbon technologies. It is only by developing and commercialising effective and affordable low carbon technologies that climate change mitigation costs can be reduced in the long term and for the world to have a realistic chance of limiting global temperature rise to around 2°C. There are other concerns around reliance on international credits from non-Annex I countries. For example, there is evidence[10] that some of the emissions reductions are non-additional (ie they would have occurred anyway), resulting in a net increase in Annex I GHG levels.

---

8 Eliasch Review modelling, including the Office of Climate Change's Global Carbon Finance (GLOCAF) model.
9 The EUA price is also referred to in this Review as the EU carbon market price.
10 For example one report estimates that approximately 20 per cent of CDM credits were likely not to be additional in the period studied: Oeko-Institut (2007)

This Review commissioned modelling to show the impact of adjusting two variables on the EU ETS price during Phase III[11] of its operation: the level of the supplementarity limit and emissions reduction target.[12] The EU ETS was chosen for the modelling as it currently accounts for the majority of the global carbon market. This piece of work modelled the international credit market without the inclusion of forest credits. The scenarios modelled are illustrated in Figure 11.3 below.

Figure 11.3: Illustration of the modelling of EUA price for a given emissions reduction scenario against different supplementarity levels (excluding forest credits)

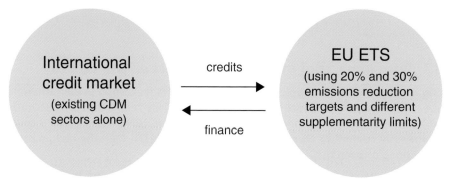

The results of the modelling are presented in Table 11.1 below. The projected EU Allowance prices (EUAs) are given using indices, with the result for the 30 per cent supplementarity limit in the 20 per cent emissions reduction scenario being used to give the reference value of 100. The results demonstrate that the higher the supplementarity limit, the lower the EUA price. This is illustrated by Diagrams A and B in Figure 11.5 below.

The modelling results also show that if the supplementarity limit is raised simultaneously with increasing the stringency of the emissions target then the EUA price need not be reduced. For example, Table 11.1 shows that the EUA price in Phase III is similar whether: (a) emissions are cut by 20 per cent and the supplementarity limit is 30 per cent or (b) emissions are cut by 30 per cent and the supplementarity limit is 50 per cent. In the second scenario the EU would be cutting emissions more and driving more abatement in developing countries without increasing the costs to European industry.

---

11 The compliance period that is due to commence at the end of the current period, which runs to 2012.
12 The emissions reduction target levels modelled (20 and 30 per cent cuts) are those proposed by the European Commission for Phase III of the EU ETS. The emissions cuts are relative to 1990 levels.

Table 11.1: Impact on Phase III EUA price of different supplementarity limits for international credits (excluding forestry) and against different EU emissions cuts

EU supplementarity limits	EUA price (indexed)	
	20% EU reduction target	30% EU reduction target
15	115	141
30	**100**	123
50	76	97
70	46	62
85	31	45
100	22	41

Source: Modelling for the Eliasch Review

11.3.2 Price impacts of forest credits on the EU ETS

Modelling suggests that if supplementarity limits are set at 50 per cent or lower in Phase III of the EU ETS, then admitting forest credits into the international credit market would have little or no impact on the EU carbon market price. This is because when restrictions on the use of non-Annex I country credits are this tight, more costly EU abatement would still be necessary and would continue to set the price for all units of abatement in the carbon market.

To study the impact on the EU carbon price of introducing forest credits to the international credit market, the modelling described in Section 11.3.1 above was repeated with the only difference being that the international credit market now also contained forest credits. The scenarios modelled are illustrated in Figure 11.4 below.

Figure 11.4: Illustration of the modelling of EUA price for a given emissions reduction scenario against different supplementarity limits (including forest credits)

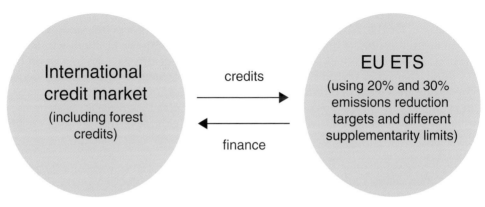

The results of this second series of modelling runs are presented in Table 11.2 below. The same reference value is used to present EUA prices in Table 11.2 as was used for the indexing that produced the indexed EUA prices in Table 11.1. The results from this second round of modelling show that in the 20 per cent EU emissions reduction scenario, inclusion of forestry has no impact if supplementarity limits are already set at 50 per cent or lower. In the 30 per cent EU emissions reduction scenario, no EUA price impact is detected until a supplementarity limit of 85 per cent is reached.

Table 11.2: Impact on the Phase III EUA price of adding forestry credits to the international credit market

EU supplementarity limit	20% EU reduction target		30% EU reduction target	
	EUA price (indexed)	change in price on admitting forest credits	EUA price (indexed)	change in price on admitting forest credits
15	115	0%	141	0%
30	**100**	0%	123	0%
50	76	0%	97	0%
70	45	-1%	62	0%
85	28	-9%	39	-13%
100	21	-4%	24	-41%

Source: Modelling for the Eliasch Review

At the 50 per cent supplementarity limit, the driver of EUA price change is still the supplementarity limit for all international credits – not whether forestry credits form part of the international credit market. It is only if the supplementarity limit for all credits is raised very high that the inclusion of forest credits would have an impact on the EUA price. At the time of writing the European Commission is proposing a supplementarity limit of around 35 per cent for Phase III of the EU ETS, which would be considerably lower than the 50 per cent level.[13]

The reason for this result is that the EUA price is determined by the marginal unit of abatement. It is only at high levels of supplementarity, with the EU buying in 50 per cent of its abatement from outside its borders, that forest credits have an impact on the EUA price. Abatement within the EU tends to be more expensive than buying credits representing abatement in non-Annex I countries.

It is only at very high supplementarity limits, when more expensive EU abatement is no longer needed to meet EU emissions targets, that the price of international credits

---

13 This is based upon effort in the EU ETS against business as usual emissions for the period 2008-2020, adjusted for estimated savings from the EU's renewable and energy efficiency targets. It includes effort from aviation. A recent non-paper from the European Commission suggests that 45 per cent of effort against 2005 emissions can be achieved through international credits in the period 2008-2020 (this figure excludes aviation entering the EU ETS).

becomes the marginal unit of abatement and hence the EUA price. Only at this point would adding forest credits to the international credit market lower the EUA price because (at lower levels of abatement) forest credits are relatively cheap to supply.

Figure 11.5 below illustrates this effect. The supplementarity limit is represented by C_L and the EU ETS emissions reduction target represented by T_E. Diagrams A and C show that at low supplementarity levels the EUA price (P_{EUA}) remains the same despite the international credit price (P_{CER}) being reduced by the admission of forestry to the international credit market. Diagrams C and D show that it is only when supplementarity limits become very high (without a corresponding increase in the emissions reduction target) that the admission of forestry to the international credit market starts to impact on the EUA price.

Figure 11.5: Illustration of why the Phase III EUA price is unaffected by the admission of forest credits, except at very high supplementarity levels

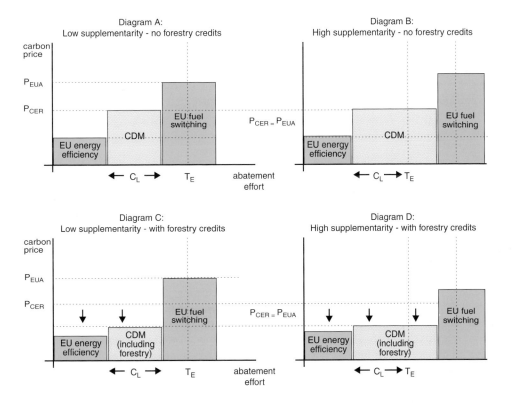

11.3.3 Impacts on the international credit market

If the forest sector is able to realise a significant amount of its abatement potential, then modelling suggests that forest credits could constitute at most 34 per cent of the international credit market in 2020.

The international community needs to balance emissions targets and supplementarity limits, in order to achieve the desired trade-off between four objectives:

- fund significant forest abatement;
- reduce the cost to industry of meeting more stringent global emissions targets;
- provide a strong incentive to invest in new clean energy technologies;
- support high level of technology transfer to the developing world.

Admitting forest credits to the international credit market without also increasing the stringency of emissions targets or raising supplementarity limits would affect the composition of that market. It is important to note that any impact on CDM sectors would not reduce the incentive to develop new clean energy technologies. Those technologies are principally developed for and incentivised by economies that place tight limits on their GHGs, which are by definition not those supplying CDM credits. However, if displacement of some of the CDM power sector were to occur, there could be an impact on the uptake of cleaner technologies in developing countries with the potential consequences highlighted at the start of Section 11.3.

The Eliasch Review modelled the impact of the admission of forest credits on the potential composition of the international credit market. Demand from all carbon markets – not only the EU ETS – was included. Three key assumptions were used to model the largest possible impact forest credits could have on the international credit market. These assumptions are that on the admission of forest credits:

- global emissions target remains the same;
- supplementarity limits remain the same;
- the forest sector is able to realise a relatively large proportion of its potential abatement compared to other sectors (despite the capacity constraints faced by the forest sector – see Chapter 13). The CDM has to date only been able to realise a limited amount of the potential abatement in its sectors. In our modelling, however, we have assumed that forestry is able to realise more of its potential than in other sectors.

Bearing in mind that the scenario modelled would be unlikely to occur in practice, the modelling results from this scenario suggest that allowing forest credits into the international credit market could generate around 600 $MtCO_2$ abatement from the forest sector in 2020. This abatement would displace CDM credits from other sectors only if the total global abatement target remains the same. Under these conditions, the forest sector could constitute up to 34 per cent of the international credit market in 2020 (see Figure 11.6). Although this shows that forestry would not completely crowd out other sectors, it would have an impact on the finance flows for technology transfer. However, if the global abatement target was raised (which admitting the forest sector would make more affordable) and/or supplementarity restrictions eased, then finance flows for both technology transfer and forestry abatement would be higher.

Figure 11.6: Top-end estimate of the share of forestry abatement in the international credit market in 2020

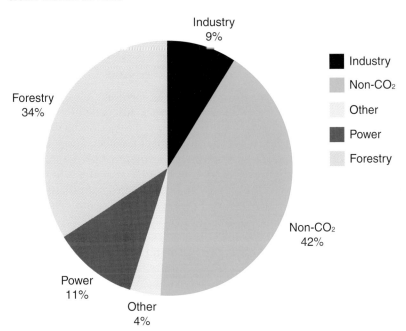

Source: Modelling for the Eliasch Review

Further modelling was then undertaken to show what proportion of total global abatement (not simply abatement in the non-Annex I international credit market) this would constitute. The share of 34 per cent of the international credit market translates to around 9 per cent of total global emissions abatement in 2020 (see Figure 11.7 below). Even though this is a high end estimate because of the assumptions used, this share of global abatement is relatively small considering forest sector emissions constitute around 17 per cent of global GHG emissions. The key reason for this is the imposition of supplementarity limits by Annex I countries.

Figure 11.7: Top-end estimate of amount of forest sector abatement as a proportion of total global abatement in 2020

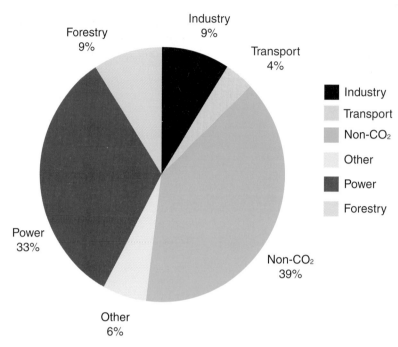

Source: Modelling for the Eliasch Review

11.4 Scale of carbon market finance for forest abatement

The choices the international community make on the stringency of global emissions targets, supplementarity limits and credit design features will have a large impact on the amount of carbon market finance available for reducing forest emissions.

One scenario modelled suggests that the global carbon market could supply around $7 billion per year for forest abatement in 2020. This would leave a funding gap of around $11–19 billion for halving forest emissions, which would need to come from other private and public sources.

Under the core scenario modelled[14] for this Review, and described below, carbon markets would generate $7 billion per year for forest emissions abatement in 2020, which would be sufficient to reduce deforestation emissions by 22 per cent relative to BAU. The modelling suggests that around $11–19 billion per year of further funding would be needed from other sources to halve forest emissions in 2020 (see Figure 11.8). Using the same

---

14 By the UK Office of Climate Change's Global Carbon Finance (GLOCAF) model, using the 2020 IIASA marginal abatement cost curve (MACC) commissioned for this Review (see Chapter 5).

core scenario assumptions, the modelling results suggest that by 2030, 75 per cent of all potential forest abatement could be financed from a global carbon market.

The $11–19 billion per year figure is given as a range because it could include up to $8 billion of rent, which it may or may not be appropriate to finance from public funds.[15] Forest protection policy costs are not included in this range. Furthermore, the modelling includes emissions reductions from reduced deforestation and degradation (REDD) but not from afforestation, reforestation and restoration (ARR).

Figure 11.8: Modelling scenario showing finance from carbon markets and other sources

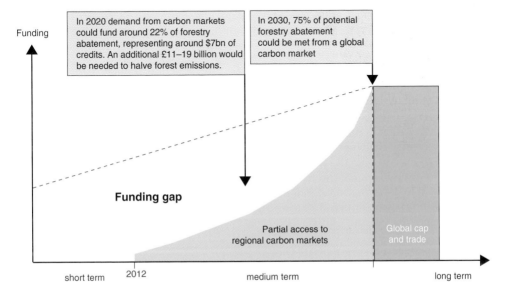

The key drivers of the level of carbon market finance available for forest abatement are: the overall demand for international credits, the rules of the system for generating forest credits and the relative price of forest credits compared to those from other sectors. In particular, these key drivers are:

a) The global emissions target: the tighter the overall target, the greater the demand for forest credits.

b) Supplementarity limits: the higher Annex I supplementarity limits, the greater the demand for forest credits.

c) The supply of forest credits: national baseline-credits systems, as recommended by this Review, are likely to generate more credits than project based schemes (see Chapter 9).

d) The supply of credits from other sectors: the entry of relatively cheaper credits from other sectors would tend to crowd out forest credits.

These drivers were modelled by this Review as follows to give the funding gap figure of around $11–19 billion per year in 2020:

---

15 For a discussion of the rent issue, see Chapter 5.

a) The global emissions target: the world aims for a 475ppm CO_2e stabilisation trajectory, initially overshooting to 500ppm. This means that global GHG emissions in 2020 are 30 per cent higher than in 1990.
b) Supplementarity limits: the supplementarity assumptions are based on the European Commission's current proposals for Phase III of the EU ETS and on proposals contained in the Warner-Leiberman Bill for the US. For all other Annex I countries that have targets, we have assumed 50 per cent supplementarity limits.
c) The supply of forest credits: a relatively efficient national baseline-credit system is assumed.
d) The supply of credits from other sectors: CDM efficiency is assumed to improve so its transaction costs are lower, and it is widened to encompass more sectors. But it is assumed that in 2020 there are still significant inefficiencies.

Any changes to these assumptions would alter the results presented above on the level of forest abatement finance that carbon markets could generate in 2020. The choices the international community make on the stringency of emissions targets, supplementarity limits and other design issues will impact on the amount of carbon market finance available for forest abatement. If relatively low supplementarity levels are applied without much tighter Annex I emissions targets, then more non-carbon market funding will be needed for forestry abatement. Chapter 13 discusses the other potential sources of finance that could be used to plug the funding gap, and uses the $11–19 billion per year figure to give an idea of how much extra funding might be needed in the medium term. Section 11.5 below discusses how the different sources of funding might be best combined for the purchase of forest credits.

11.5 Linking mechanism

In Section 11.4 we conclude that in the transition to a comprehensive global cap and trade system, funding for forest abatement is likely to need to come from both the carbon market and other private and public sources. This raises the question of whether and how to combine different funding sources to purchase forest credits from forest nations.

In this section we examine three important functions that a linking mechanism, or institution will need to provide:

- aggregation of different funding sources;
- managing the risk of reversal of emissions reductions;
- reducing the risk of investment for forest nations.

In practice, it is likely that an institution will be needed in order to perform these linking functions.

11.5.1 Aggregation of different funding sources

A single institution should be used to aggregate different sources of funding and purchase all the forest credits generated by forest nations.

A single forest credit purchasing mechanism would be able to aggregate funding from carbon markets and other sources. Having separate purchasing mechanisms for different sources of funding would be inefficient and would result in unnecessary duplication and complexity.

Moreover, different purchasing mechanisms representing different sources of non-market funding would be likely to offer different prices for forest credits. Some funders might have the power to fix prices at a certain level. If prices were different then this would threaten to undermine the realisation of potential forest abatement. Forest nations could experience a perverse incentive to hold off from making emissions reductions until, say, a non-market fund offering a very high price for credits was expected to become established.

For these reasons, this Review recommends that there should be just one forest credit purchasing institution that aggregates funds from the various different sources. This is illustrated in Figure 11.9.

Figure 11.9: Aggregation of funding sources

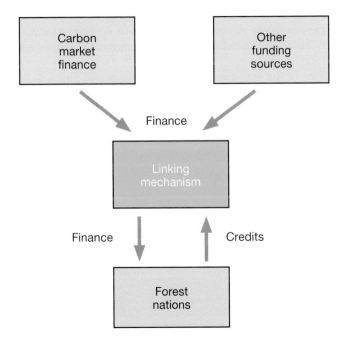

11.5.2 Managing the risk of reversal of emissions reductions

A proportion of the forest credits generated should be placed within a reserve and used to offset any future reversal of forest emissions reductions.

In all sectors, a fall in emissions in one period might be reversed in the next. In the energy sector this might occur, for instance, as a result of a cold winter. In the forest sector a reversal might occur as a result of increased illegal conversion of forest land as a result of a spike in agricultural commodity prices.

The first point to note is that forest emissions arise predominantly from non-Annex I countries. There are currently no caps on these countries' emissions, not only for forest emissions but all sectors. In a comprehensive global cap and trade system, permanence would be assured by financial penalties for exceeding a country's GHG emissions cap (as is currently the case for Annex I countries). However, even without a binding emissions cap there are practical constraints on a country reversing its emissions reductions. To take the forest sector as an example, a forest nation that has invested money in reducing forest emissions is unlikely to want to allow emissions to rise as it would not be able to recoup its investment. Furthermore, a certain amount of infrastructure is needed to convert forest land to productive farmland.

One peculiarity of the forest sector is that it can act as a sink as well as a source of emissions. Credits can currently be awarded for project-level afforestation and reforestation (A/R) under CDM. These credits would be awarded for carbon sequestration: the carbon absorbed and stored in trees as they grow. It may take 100 years for a plantation to reach maturity, and during that time the carbon might be released due to fire, disease or logging. This would result in the sequestration that has been credited and paid for in advance of maturity not in fact taking place.

This is a significant problem for CDM A/R projects, where a large proportion of the forest carbon could be lost in any one event. Rules were therefore developed for CDM A/R projects to address the risk of impermanence: any credits issued would expire after a period of time. Temporary CERs (tCERs) expire at the end of the commitment period subsequent to the one in which they are issued, while 'long-term' CERs (lCERs) expire at the end of the project period. Neither type of CDM A/R credit can be carried over to subsequent commitment periods. Furthermore, responsibility for lCER replacement in the event of a reversal of removals falls to the purchaser who has retired the credits.

These rules have made CDM A/R credits a substantially lower class of credit than CDM credits from other sectors: they are estimated to be worth around 25 per cent of the value of standard CERs.[16] This unattractiveness to potential purchasers is partly responsible for the fact that only one CDM A/R project has been registered at the time of writing.

Reversals of emissions reductions is less of a problem for the forest sector at the national level because in most cases any net emissions reductions are likely to involve only a small proportion of sequestration relative to reduced deforestation and degradation emissions. Furthermore, at a national scale, it is much less likely that a large proportion of forest carbon would be affected by any one-off event in a given period. This is one of the reasons why a national-level approach to forest abatement is recommended in Chapter 9.

---

16 Baalman and Schlamadinger (2008)

Even if there were a long-term liability for forest carbon stocks, which is one of the proposals under discussion in UNFCCC negotiations, the risk associated with a reversal of net forest emissions above a reference level would need to be managed in a manner that ensures:

a) environmental effectiveness and credibility;
b) little or no risk to credit purchasers;
b) sufficient stability for forest nations to pursue effective emission reduction policies

A means of meeting all three criteria is through the creation of a forest carbon reserve (or buffer) at the national level, following an approach used for the project level in the Voluntary Carbon Standard (VCS). This would involve setting aside a percentage of the credits awarded to a forest nation in a reserve account. Should the nation's emissions rise rather than fall relative to its reference level in a successive period then a quantity of credits sufficient to cover the reversal would be cancelled out.

To further spread the risk of reversal, national credit reserves could be aggregated into one central reserve. This already occurs between projects in the VCS scheme. The risk of reversal would be spread between all countries. At a global level, the deforestation rate remains fairly stable. If one country lost a significant proportion of its forest then this might cause a large increase in forest emissions nationally, yet become insignificant at the global level. The reversal in the country's emissions might be larger than the quantity of carbon stored in its reserve account, so the country would be able to use credits from the central reserve to make good the whole reversal. The more countries participating in national baseline-credit schemes, then the more effectively, efficiently and equitably risk would be spread. Figure 11.10, illustrates how the central credit reserve scheme might work.

Figure 11.10: Illustration of a credit reserve for managing the risk of reversal of forest emissions reductions

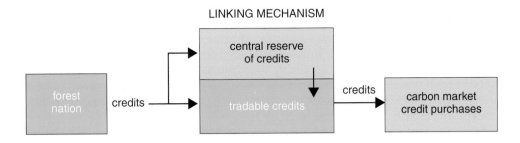

If a central reserve is used, then a system of incentives would need to be adopted to deal with the risk of moral hazard. If a forest nation thought that it could rely on other countries' reserve credits to make good its own shortfall, then it might not make a real effort to prevent the reversal from occurring. The risk is addressed to a certain extent through credits being awarded ex poste: the country needs to ensure successive emissions reductions to secure any payment at all. The moral hazard could be further addressed by a rule which required a proportion, for example 10 per cent, of a countries' credits, to go into a national reserve and a further a proportion, for example 10 per cent, to go into a central reserve. If, at the end of the crediting period, the nation's central reserve credits were not called upon then it could be allowed to sell the residue of credits in its national reserve.

A credit reserve system with a sufficiently large reserve to cover the risk of reversals would generate credits that were attractive to purchasers and comparable to other CDM credits. But the system would have to be designed carefully. Initially unknown and then changing reserve allocations would have an adverse impact on the incentive to invest in forest emissions reductions. The allocation will need to be set sufficiently high for the reserve to be large enough to cover any emissions reversals, yet not so high as to disincentivise investment in emissions reductions. The credit reserve could be further enhanced with credits that are purchased from the aggregated fund but are not destined for resale to carbon markets. Taken together with the reserve allocation, this should ensure a sufficiently large reserve. Even in the case of an agreement on forestry that involves long-term liability for reversals, a credit reserve system could help to smooth out fluctuations in deforestation rates and reduce the associated compliance risks.

Another option that has been discussed for addressing the risk of reversals is that any increase in emissions in one period would lower the baseline in the successive period so in the successive period the forest nation would have to make good its reversal from the previous period before it could start to earn any credits. One problem with this option is that it weakens the incentive for the forest nation to continue to participate in a national baseline-credit scheme. The baseline in the successive period could be so challenging to reach that the nation writes off the prospect of earning further credits and decides to pull out. Other options under discussion include private insurance policies against reversals and discounting of forest sector emissions reductions to account for the risk of reversals.

11.5.3 Managing the risk of investment for forest nations

An institution could also be used to reduce the risk of investing in forest abatement for forest nations by guaranteeing to purchase from them a minimum quantity of credits at a minimum price.

A linking mechanism could also be used to minimise the risk to forest nations of investing scarce financial resources in forest emissions reduction policies and for the international credit market price then to fall. In the absence of intervention, the credit price would fluctuate in line with fluctuating supply and demand for international credits. This is the normal function of markets and the means by which an efficient credit price is determined. However, some forest nations may not be well suited to making substantial investments in emissions reduction policies in such circumstances.

A linking mechanism could therefore guarantee forest nations a minimum purchase price for their forest credits and also a minimum quantity of credits that it would purchase in a given period. This would give forest nations the certainty they need to justify investing scarce resources in forest emissions reduction policies. Still, the international credit market price is just one factor that determines the quantity of credits supplied to the market in any given period. Other factors include the credit reference level (see Chapter 9), the forest nation capacity for pursuing emissions reduction policies and access to up front finance (see Chapter 13).

In setting a single guaranteed price, a trade-off would need to be made between on the one hand ensuring equity and strong incentives to forest nations and, on the other hand, ensuring that the forest sector's low-cost abatement potential is not negated. It would also be important that the guarantee price does not change too frequently or unpredictably, as this will again weaken the incentive for forest nations to invest in forest emissions reduction policies.

Those forest credits that the linking mechanism did sell to the market need not be sold on immediately. This might be the case if the market price for the forest credits is currently less than the guarantee price at which they were purchased from forest nations. The linking mechanism should be permitted to temporarily hold onto the forest credits so that when they are sold they reach more than the guarantee price. This would allow it to recoup its administration expenses and to pay back any loans it took on in order to purchase credits from forest nations in the first crediting period.

However, if an institution were to play this role, discretion over the retention of forest credits, particularly if market prices are influenced, would need to be exercised carefully. If through the actions of the linking mechanism, the international credit price was artificially inflated, the benefits of low cost abatement from forestry would not be realised. It is also worth noting that there would be no particular need for the linking mechanism to temporarily restrict the supply of credits to the carbon market in order to avoid 'flooding' individual markets. This is because each carbon market is likely to have set itself a supplementarity limit sufficient to protect itself from such an eventuality (see Section 11.3).

Finally, the linking mechanism would need to operate as predictably and transparently as possible to promote stability in the markets. Stability can be enhanced in a number of ways, including by selling significant quantities of forest credits on long forward contracts shortly after they have been purchased from forest nations. A well functioning market will also require the linking mechanism to be set up in such a way that costs, delays and unpredictable discretionary decisions are kept to a minimum.

11.6 Conclusion

This chapter examined the impact of adjusting emissions reductions targets and supplementarity limits on the EU ETS carbon price. We show, with the use of modelling, that if Annex I countries simultaneously commit to more stringent emissions targets and higher supplementarity limits then they can achieve greater global emissions cuts for a given cost while maintaining incentives to invest in new low carbon technology.

This chapter also presented the results of modelling the impact of linking forestry to carbon markets on the EU ETS carbon price. The results suggest that if supplementarity limits are set at 50 per cent or lower, then admitting forest credits will have little or no impact on the EU carbon market price and therefore on the incentive to develop new low carbon technologies.

Next we modelled the potential impact of forest credits on other international credit sectors, and showed that Annex I countries will need to balance more stringent emissions targets and supplementarity limits to support the power abatement sector in developing countries, which drives low carbon technology transfer. The balancing of these two variables will also affect the amount of carbon market finance available for forestry abatement, which the results of one scenario modelled suggest could be around $7 billion in 2020.

Finally, we made the case for using a linking mechanism in the transition to a comprehensive global cap and trade system in order to blend carbon market finance with other sources for purchasing credits. We recommend that the linking mechanism should be used to manage the risk of reversal of forest emissions reductions as well as to reduce the risks to forest nations investing in emissions reduction policies.

The next chapter explores the governance requirements necessary for forest nations to respond to the incentive provided by carbon markets and other funding so that they can realise a significant amount of the world's forest abatement potential.

12. Governance and distribution of finance

Key messages

Good governance and effective mechanisms for the distribution of finance are needed for a successful framework to reduce forest carbon emissions.

National governments need to take the lead in implementing a successful system to tackle deforestation. Key areas of governance reform include clarifying and securing land tenure and user rights and strengthening the institutional capacity of national, regional and local institutions. Determining and implementing changes with the full participation of forest communities will make reforms more likely to succeed and benefit the poor.

National governments should decide the level of delegation for both the implementation of policies and programmes to reduce deforestation, and the receipt and use of revenue from forest credits.

Many policy and programme options exist for reducing emissions from deforestation that do not require cash transfers to individuals. However some options will do so, including transfers to subsistence farmers and foresters. Such delegation will involve significant costs and capacity requirements, which may be very challenging for many forest nations in the short term. Capacity building and demonstration activities to test these approaches will be needed.

To help promote transparency, accountability and strong financial management, countries may choose to manage carbon revenues through a special fund with broad-based participation in its governance, which could include international institutions.

Countries participating in a forestry mechanism should report on the policies and measures they have put in place to reduce deforestation, including an assessment of social and environmental impacts, and information on stakeholder consultation processes. These communications should be subject to peer review.

Forest nations should also consider establishing a transparency mechanism for forest carbon finance transfers, and the international community should provide appropriate support.

Programmes to reduce forest loss should avoid unintended harm to the people and places they affect. In addition, voluntary higher standards for programmes that aim to achieve wider benefits such as poverty reduction and the protection of biodiversity could be created. Premium credits generated from such programmes could be given preferential treatment in the international compliance or other markets.

12.1 Introduction

Good governance and effective mechanisms for the distribution of finance are needed for a successful framework to reduce forest carbon emissions.

Achieving any revenue from a forestry mechanism depends on countries being able to control and manage their forest resources. Implementation may be through decentralised programmes, involving many different stakeholders and with varying levels of central government involvement. Sub-national approaches will also be important in continued piloting and for engaging communities and the private sector. This chapter examines sub-national delegation and distribution systems, and the challenges these imply. Given that solutions will necessarily be country-specific and often implemented at the local level, this chapter seeks to set out options, general principles and challenges rather than making specific recommendations about which policies, measures and systems would be most appropriate or successful in particular forest nations.

The chapter also addresses the role of the international system in ensuring basic social and environmental safeguards for the people and places affected by new policies and measures to reduce deforestation, and in promoting best practice.

12.2 National-level governance

National governments need to take the lead in implementing a successful system to tackle deforestation. Key areas of governance reform include clarifying and securing land tenure and user rights and strengthening the institutional capacity of national, regional and local institutions. Determining and implementing changes with the full participation of forest communities will make reforms more likely to succeed and benefit the poor.

An international mechanism on forestry must respect the sovereignty of nation states and their rights to determine the best uses of their forests for the benefit of their citizens. The main role of national governments is in developing and implementing an overall strategy for reducing deforestation consistent with broader national development objectives. Drivers of deforestation vary widely between and within countries, so solutions need to be country-specific and based on country-led analysis of the economic, social and institutional drivers of forest loss.

National strategies and mainstreaming

Climate change, deforestation, sustainable development and livelihood considerations need to be integrated into national growth and development strategies. It will be necessary to identify and remove inappropriate financial incentives for forest conversion, where they exist, and similarly to reform tax/subsidy regimes to incentivise forest protection. The effectiveness of policies to reduce deforestation depends on policies and action in other sectors. Systematic application of environmental and social impact assessments to all major policy developments will be the key means for governments to expose the inevitable trade-offs between different policy objectives, make decisions in the full knowledge of the likely impact on deforestation and rural livelihoods, and put in place mitigation strategies where necessary.

The prospect of a new international mechanism on forestry provides an incentive for rainforest nations to reassess and revise their legal and policy frameworks to ensure a) their appropriateness for enabling a country to benefit from a future scheme; and b) clarity of coverage and application. Box 12.1 describes Guyana's low carbon development vision, which aims to transform the country's economy and achieve its development goals while preserving its natural resources.

> **Box 12.1: Guyana's low carbon development vision**[1]
>
> The Government of Guyana is working to produce a low-carbon development strategy which identifies how to:
>
> - harness low carbon opportunities to stimulate job creation, investment and economic growth in Guyana;
> - work with the global community to create financial incentives to make it more valuable to leave Guyana's 16 million hectares of rainforests standing than to cut them down;
> - protect Guyana's people and productive land from climate change – in 2005, floods caused damage equivalent to 60 per cent of Guyana's GDP.
>
> The strategy is grounded in the view that it is not necessary to choose between national development and tackling climate change, but instead to look at how to forge successful economies that avoid the high carbon development of the past.
>
> The Government of Guyana will facilitate a national consultation on the development strategy, to reach a broad national consensus on how to achieve the vision.

Land tenure and user rights
Clarifying and securing land tenure and user rights should be a priority for forest nations, for social as well as environmental reasons. Only when property rights are secure, on paper and in practice, are longer-term investments in sustainable management made possible. Clarifying and securing land tenure has evident benefits for the poor, enshrining in law their traditional rights to draw economic, social and cultural value from forests. Without clear land rights – defining who owns land or the resources on it and who can receive the income it generates – a forestry finance mechanism may prove to be high risk for the poor.[2] There will be a danger of customary rights being violated in the interests of inward investment, and abusive contracts and land speculation acting to the detriment of community interests.[3] Thus without clear tenure and use rights, sustainable forest management will be impossible and carbon finance may increase social conflict.

Land-use planning and zoning
A system for ensuring rational land use planning at national and decentralised levels will be critical. This may include new zoning programmes, undertaken in the context of the national strategy for reducing deforestation, which, for example, establish new areas of permanent forest reserve out of areas currently identified for extraction and conversion uses. All zoning exercises have implications for forest communities and existing land holders, and ensuring participation and buy-in from all interested parties will be critical

---

1   Government of Guyana (2008)
2   Peskett and Harkin (2007)
3   Griffiths (2007)

to their success. There are a range of examples from Latin America and South East Asia where zoning exercises have failed due to lack of popular support and consultation.[4] Box 12.2 describes a participatory mapping exercise in the Democratic Republic of Congo intended to provide the basis for an inclusive future zoning exercise.

> ### Box 12.2: Community mapping in the Democratic Republic of Congo
>
> The Rainforest Foundation is supporting a group of Congolese NGOs led by CENADEP to undertake a community mapping project in several locations across DRC. The team working on this Review met representatives of the project who had just returned from mapping an entire territory in DRC's Bandundu province.
>
> Community members are trained in the use of GPS to produce a map of their local area. An example is shown below. The exercise in Bandundu identified 190 villages where only 30 had been shown on a previous map. It also identified agricultural land used by communities, sites of cultural or religious significance, and the extent and limits of commercial logging and other concessions. The results were presented to a mapping workshop with the DRC government to demonstrate that large scale participatory mapping was possible, and should be the precursor to any new zoning exercise. Experience from an earlier project site, where communities successfully lobbied logging companies to provide additional infrastructure on the basis of new information gleaned from the mapping exercise, demonstrated the potential of this approach for strengthening the rights of forest communities.
>
>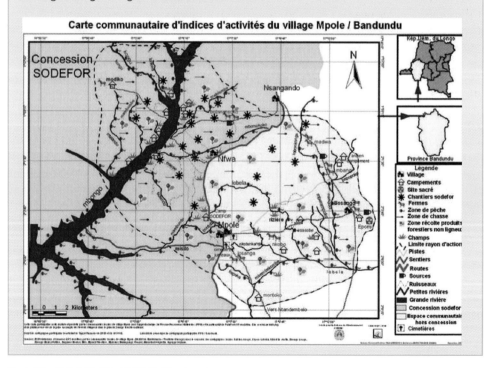

---

4   Chomitz et al (2006)

Institutional capacity
In addition to clarifying land rights and ensuring rational land use planning, the institutional capacity of national, regional and local institutions needs to be strengthened. Governments will need to ensure that different agencies have clear responsibilities and are working in concert to achieve reduced deforestation. Governments need to be able to develop, implement and enforce laws and policies that will impact on deforestation and help reduce poverty. They will also need to develop the institutions and systems required to access a new international finance mechanism, for example measuring and monitoring capability, internal transfer mechanisms and revenue collection channels for carbon finance. A country's success in reducing deforestation depends on institutions and sectors well beyond forestry and wider land use, so forestry-related reforms need to be integrated into wider processes of institutional development.

Participatory approaches
For governance reforms to benefit the poor, they need to be determined and implemented in participation with forest communities. Genuinely participatory land-use planning at all levels, with the establishment of a national consensus on forest policy through comprehensive stakeholder engagement, will be the cornerstone of successful forest governance.[5] The National Forest Programme approach[6] and the FLEGT process, which require a multi-stakeholder consultation and consensus on a definition of legality, are good models that can contribute to higher levels of trust between governments, the private sector, NGOs and community groups. Articles 18 and 19 of the United Nations Declaration on the Rights of Indigenous Peoples[7] concern the rights of indigenous peoples to participate in matters which would affect their rights, and the need for States to consult with indigenous peoples on matters affecting them. There will always be trade-offs between speed, simplicity and scalability of policy and programme development and implementation, and how closely involved all stakeholders can be. But the environmental and social sustainability of policies to reduce deforestation will depend on the buy-in of all interested stakeholders, and of those who live in and around forests in particular.

Immediate action
Improving forest governance in the context of international action on climate change is the principal means by which nation states can reduce deforestation. But action to reduce deforestation cannot wait for perfect forest governance. Instead, the development of mechanisms and demonstration activities at national and sub-national-level can be used as a platform to improve governance in parallel, through developing institutional and human capacity. Sub-national programmes can help generate locally determined reforms and agreements and do not have to wait for national systems to be in place first. New finance for forests could act as an entry point for wider governance reform in many developing countries.[8]

Chapter 13 looks at how the international community, through financial and technical assistance, could support the kind of reforms described above that will allow countries to take advantage of an international incentive mechanism for forestry.

---

5   Saunders et al (2008)
6   http://www.nfp-facility.org/home/en/
7   Resolution 61/295, adopted by the General Assembly in September 2007
8   Brown et al (2002)

12.3 Distribution of finance

It is often assumed that linking carbon finance to forests would require forest nations to make cash reward payments to large numbers of individual landholders. The logistical complexity and high transaction costs of establishing this kind of distribution system, particularly in developing countries, has been the cause of scepticism about the feasibility of an international finance mechanism for forests. In fact there are many different ways that forest nations can try to reduce deforestation, and a wide variety of uses and distribution mechanisms for the carbon finance generated. PES (Payment for Ecosystem Services) schemes involving many individual landowners is only one of the options, and countries will need to carry out their own analytical work to determine the best solution for them.

12.3.1 Sources of funding and spending options

The relationship between sources of funding, what it is spent on and the ultimate recipients is complex, will differ between countries and will change over time as new sources of funding come on stream.

As discussed in Chapter 13, funding for reducing deforestation can come from a variety of sources (see Figure 12.1). Countries could use their own funds to implement programmes to reduce deforestation, and use the revenues gained from the carbon market to finance other policy priorities. Alternatively, bilateral and multilateral funds may be used in the early years to finance policy and institutional reforms and some initial financial incentive schemes, to be replaced by carbon finance once initial emissions reductions have been achieved and forestry has been integrated into the carbon market. Whilst funds from bilateral and multilateral donors are likely to be tied to specific programmes and policies, funds from the carbon market will not necessarily be spent on programmes related to climate change and deforestation.

Figure 12.1: Different funding sources and spending options available to forest nations

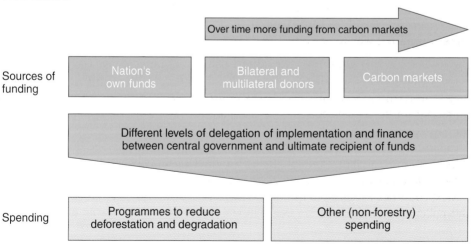

12.3.2 Government policies and measures to reduce deforestation

Forest nations have three main options for policies and measures to reduce deforestation:

- *Command and control measures.* These could include direct enforcement of forest laws, revised if necessary, and the establishment and policing of protected areas. Indirect regulation will also be important (for example, limiting the deforestation impacts of infrastructure). Command and control measures are likely to be a significant part of all forest nations' attempts to reduce deforestation.
- *Taxation.* Countries may discourage activity that leads to deforestation (such as forest clearance for agriculture) by imposing higher taxes on such activities.
- *Financial transfers.* Countries may make cash or in-kind transfers to actors who are reducing deforestation or undertaking forestation. To incentivise positive behaviour change, financial rewards could be made to actors who, for example, move to a system of sustainable forest management (SFM), or plant trees.[9]

Most governments are likely to employ a combination of command and control, taxation and financial incentives. For example, if a new piece of legislation such as the establishment of a new protected area means that certain people are no longer able to access the area's resources, they may require compensation for income forgone.

12.3.3 Delegation of programmes and finance to sub-national actors

National governments should decide the level of delegation for both the implementation of policies and programmes to reduce deforestation, and the receipt and use of revenue from forest credits.

Many policy and programme options exist for reducing emissions from deforestation that do not require cash transfers to individuals. However some options will do so, including transfers to subsistence farmers and foresters. Such delegation will involve significant costs and capacity requirements, which may be very challenging for many forest nations in the short term. Capacity building and demonstration activities to test these approaches will be needed.

The role of national governments in direct delivery of programmes on the ground to reduce deforestation – whether they be command and control or transfer measures – will vary from country to country. Central government's main role will be in creating the conditions for implementing actors (ie, sub-national state structures, private companies, communities) to successfully enact policies and programmes to reduce deforestation.

Countries will make decisions depending on national circumstances about the level of delegation of responsibility and incentive for a) implementing policies and programmes to reduce deforestation; and b) receiving and spending revenue from forest credits.

Most countries will need to introduce national-level reforms to align better their legal and policy frameworks with an international incentive mechanism for forestry. Coun-

---

9   Peskett et al (2008)

tries may choose to direct some programmes centrally – for example, the management of protected areas, putting in place support services for the promotion of SFM, or centrally managed PES schemes (see Box 12.3 for an example from Costa Rica).

Box 12.3: Costa Rica PES scheme

The Costa Rica Payment for Environmental Services scheme is an example of a successful nationwide PES scheme, involving payments to individual landowners. Landholders volunteer to participate and undertake to preserve various environmental services – including forest cover – through sustainable forest management, reforestation and restoration processes and agroforestry.

The funding for the scheme comes from tradable offsets sold on international markets, donor funds (including a Global Environment Facility grant) and a national fuel tax.

Figure A below shows the organisational structure of the system. FONAFIFO, part of the Ministry of Environment, is the organising institution responsible for reviewing applications, conducting verifications, making payments and monitoring programmes. The Joint Implementation Office is the channel through which carbon credits are channelled. Landholders can have bilateral contracts with credit buyers, but all credits must be registered centrally.

Between 1997 and 2005, half a million hectares of land had been covered by environmental service payments, the majority for forest protection, and $120 million has been delivered in ecosystem service payments.[10]

Figure A: Organisational structure of the Costa Rica PES scheme

10 Government of Costa Rica (2006)

Alternatively, national governments may devolve policy and implementation responsibility to *provincial, regional* or *local governments*. The Bolsa Floresta programme in Amazonas State in Brazil is one example of a provincially run PES scheme. In another example, the Aceh Green programme in Indonesia is one province's cross-sectoral response to the challenges of climate change and environmental sustainability.

National or local governments could choose to devolve responsibility for design, delivery and management of programmes to reduce deforestation to *private companies, NGOs* or *communities*. Whichever entity has ultimate control over programme design and delivery, the engagement of forest communities, and the deployment of their expertise and knowledge as those closest to forest resources, will be critical to ensure the sustainability of actions.

In a system of national targets, trends in emissions, removals and credits relative to national reference levels need to be registered nationally to avoid double counting, but devolution of responsibility for delivering programmes could be accompanied by devolved responsibility for receiving credits. Local governments, companies, communities or individual landholders could receive and manage credits, including cashing them in and investing the revenues. Figure 12.2 is a representation of this type of system.

Figure 12.2: Different options for national governments to devolve responsibility to sub-national actors

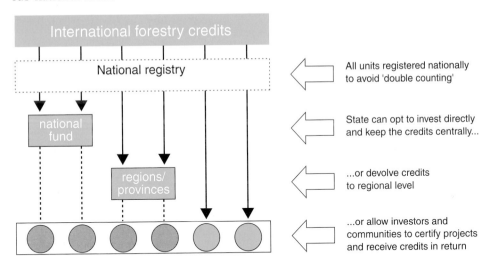

12.3.4 The challenges of delegation and distribution

Full devolution of responsibility for credits to sub-national actors, or any scheme involving distribution of cash payments to a potentially large number of individual actors, could present a significant implementation challenge for many forest nations. Work commissioned for this Review outlined the following governance requirements for large-scale delegation of responsibility for emissions reductions:

- *Clarity on resource ownership.* Where land tenure is unclear, landowners cannot be sure that their ownership or control over a project area will not be contested once it becomes more valuable. Uncertainty over whether the party selected for a carbon purchase and/

or investment contract is in fact the inalienable owner of the land/asset renders investment in land-based activities expensive and high risk.
- *Clarity on environmental service ownership/responsibilities.* Even where land tenure is certain, in some cases there are no clear legal rules for ownership of non-traditional assets, such as environmental services. There may be a need for clarification as to the legal nature of emissions reductions and, ultimately, their ownership.
- *An effective judiciary.* Related to clear land tenure, developers/investors will also need an assurance that the legal infrastructure in a country is able to uphold the rights set out in project contracts. The institutions do not necessarily need to be independent but a reasonable degree of transparency is required. The capacity to provide this contractual certainty has been noted as one of the reasons for the overwhelming dominance of China in the CDM, compared with states elsewhere in Asia and Africa.[11]

The transaction costs and capacity requirements of dealing with a large number of individual stakeholders, including the development of methodologies for calculating the precise emissions reductions that individual areas and actors are responsible for, will be challenging for many forest nations, especially because of the often highly dispersed and remote location of many people living in and near forests.

Cash/credit distribution to individual actors is only one way of reducing emissions from deforestation, however. Many of the possible policy and programme options for reducing emissions from deforestation, which nations can select according to their individual circumstance and capacity constraints, do not require cash transfers to individuals. Examples include:

- indirect or in-kind payments (for example to villages) for establishing broader development projects such as improving social services;[12]
- government-backed technical and managerial support services to SMEs to implement sustainable forestry or agricultural practices so that it becomes financially viable for them;
- broader low-carbon development strategies drawing people away from deforesting activity, for example Guyana's plans for the development of ICT parks and supporting services;[13]
- programmes to support agricultural intensification, linked to provisions for forest protection that distribute equipment and seeds, and the roll out of training programmes rather than cash;
- direct payments only to large companies (such as mining, logging, soy) to take on an environmental stewardship role in partnership with communities in the surrounding area.

Many options, however, will require direct connection with, and transfer to, many individual actors, including reaching out to highly dispersed subsistence farmers and foresters. Carbon finance may need to reach individual landholders to provide up-front funding to make the shift to sustainable practices, or as an ongoing additional income stream to make these practices economically viable. Existing local institutions, such as

---

11 Hoare et al (2008)
12 Peskett et al (2008)
13 Government of Guyana (2008)

village committees, banks and credit unions, could be used for channelling and redistributing payments, but these would need wider coverage and significant capacity building support.

National-level policy and legislative reform can take place relatively easily in capitals, but implementation and enforcement will require linkage deep into the forests. Truly participatory processes that bring forest communities into decision making also require mechanisms that can reach down to the community and individual level. This will be one of the major challenges for countries wishing to benefit from an international forestry mechanism. The spin-off benefits for other aspects of forest nations' economies and society from these developments are evident. Mobilising civil society and religious networks that generally have significantly more reach than the state will be crucial. But the challenges remain significant.

12.3.5 Management of finance by governments

To help promote transparency, accountability and strong financial management, countries may choose to manage carbon revenues through a special fund with broad-based participation in its governance, which could include international institutions.

The money that governments receive from bilateral and multilateral funds to implement programmes to reduce deforestation will be governed according to the agreement in place between the partner countries. Funds from the carbon market going directly to nation states could be managed in two different ways, outlined in Figure 12.3. This figure only analyses what government, national or regional, could do with carbon revenues. As discussed above, they would also have the flexibility to devolve responsibility for managing credits to sub-entities.

Figure 12.3: Options for managing carbon market finance

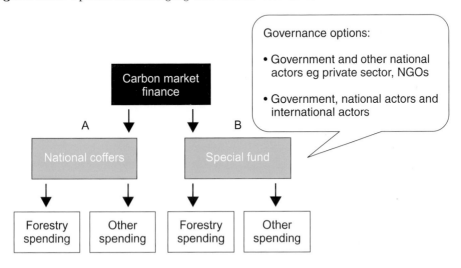

In option A above, carbon market finance is channelled directly to national governments, to use like any other sovereign revenue, on measures to reduce deforestation or otherwise.

In option B, carbon market funds from forestry are channelled into an earmarked fund. Again, they can be spent on forestry or non-forestry programmes. The main difference is in the governance of the funds. To ensure that new carbon revenues are seen to be being used in an equitable way for the benefit of all citizens, countries may choose to take a participatory management approach to the funds. This could include:

- national actors only, or
- international representation and support, such as a regional development bank to provide public financial management and capacity building support. Box 12.4 provides an example of the Kecamatan Development Fund in Indonesia, which operates in a similar way.

This kind of broad-based participation in the governance of carbon revenues would help promote transparency, accountability and strong financial management. Participation of international bodies in decisions on how to spend their funds would necessarily need to be subject to considerations of national sovereignty. Table 12.1 sets out the pros and cons of the different approaches.

Table 12.1: Pros and cons of different approaches to managing carbon revenues

	Pros	**Cons**
Option A – as other revenue	• Helps build capacity of government budgetary processes and systems • More streamlined processes • Enables holistic view and cross-government prioritisation of expenditure	• Less easy to demonstrate transparency and accountability (although delivery mechanisms could be designed to ensure this) • Capacity gaps, especially in public financial management, could hinder effectiveness of programmes
Option B1 – special fund (national actors only)	• Increased transparency and accountability • Increased confidence in equity of benefits sharing	• Potentially slower decision-making and action • Could tend to emphasise projects over a national approach

	Pros	Cons
Option B2 – special fund (with international actors)	• Increased international confidence (including from private sector) possibly leading to increased finance flows • Increased confidence in equity of benefits sharing • Access to technical assistance • Could blend market funds with those received from international public sources	• Would need to be acceptable given sovereignty considerations • Slower decision-making and action • Could tend to emphasise projects over a national approach

Box 12.4: Kecamatan Development Program: special fund mechanism[14]

The Kecamatan Development Program (KDP) is a national Government of Indonesia programme, implemented by the Ministry of Home Affairs, aimed at alleviating poverty, strengthening local government and community institutions, and improving local governance. The KDP is in its third phase, and is expected to run until 2009.

The programme is funded through government budget allocations, donor grants, and loans from the World Bank. Funds are held in a special earmarked account, independent of national budgets. The Figure below shows the fund management structure of the KDP.

It provides block grants of approximately Rp 500 million to 1.5 billion (approximately $50,000 to $150,000) to sub-districts (kecamatan) depending upon population size. Villagers engage in a participatory planning and decision-making process to allocate those resources for their self-defined development priorities. KDP focuses on Indonesia's poorest rural communities.

KDP provides funds from the national level to the village collective accounts at the kecamatan level. These are used to fund infrastructure projects, loans or social investments. Accountability and transparency are increased by each financial transaction downwards being matched by a similar paper trail upwards.

---

14 Peskett et al (2008) and www.worldbank.org/id/kdp

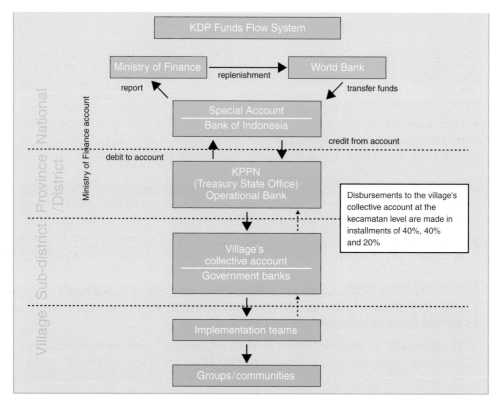

12.3.6 Capacity for distribution systems

The challenges in introducing extensive sub-national transfer and distribution systems are the same in all policies and programmes to reduce deforestation:

- governance capacity, particularly extent of state control and implementation capacity into provinces and remote areas;
- public financial management, and the existence and capacity of national and sub-national financial transfer mechanisms.

A key element of the design of national strategies to reduce deforestation, and capacity building and institutional strengthening to support their implementation, is establishing what kind of internal distribution, delegation and contractual mechanisms best suit a country's individual circumstances. The role of non-state actors, including the private sector and particularly the banking and finance sectors, will be crucial. Capacity will also need to be built outside the state. Demonstration activities will also be vital in testing the distributional aspects of certain policy approaches. Chapter 13 describes the requirements for capacity building support and demonstration activities in more detail.

12.4 International governance

National sovereignty will rightly be paramount in negotiations on bringing forestry and wider land-use issues into an international framework on climate change. Nevertheless, the international community has an important role in working with national governments to support their and their citizens' participation. This role is threefold:

- First, national level governance in many forest nations needs to be aligned and strengthened so that countries can effectively reduce deforestation. Chapter 13 looks at how the international community can best support countries to develop the systems necessary to benefit from a forestry mechanism.
- Second, the international community has a role in ensuring that basic safeguards are in place for the people and places who are affected by a new system to reduce deforestation. Policies and programmes to reduce deforestation should do no harm.
- Third, the international community can help promote best practice and policies and measures that, as well as reducing deforestation, promote co-benefits such as poverty reduction and biodiversity protection and enhancement.

While carbon finance for forestry could potentially provide significant additional income to some of the poorest countries in the world, concerns have been raised about its impact in forest nations if administered poorly. Concerns include:

- The impact of potentially large new income streams in countries where governance and public financial management are weak or respect for human rights is poor.
- The impact on indigenous peoples and forest communities, who often experience vulnerability and marginalisation. Risks include:
  - renewed or increased state and outside control over forests
  - support for exclusionary models of forest conservation
  - violations of customary land and territorial rights
  - unequal and abusive community contracts
  - land speculation, land grabbing and land conflicts (competing claims for compensation for avoiding deforestation).[15]
- The implications for associated issues such as biodiversity protection.

This Review has identified two key ways in which statutory and voluntary governance of the international climate change system can help address some of these concerns:

1. national reporting requirements and transparency mechanisms;
2. voluntary standards to promote best practice.

12.4.1 Reporting and transparency

Countries participating in a forestry mechanism should report on the policies and measures they have put in place to reduce deforestation, including an assessment of social and environmental impacts, and information on stakeholder consultation processes. These communications should be subject to peer review.

---

15 Griffiths (2007)

Forest nations should also consider establishing a transparency mechanism for forest carbon finance transfers, and the international community should provide appropriate support.

Reporting requirements
Reporting to the UNFCCC should become the key tool for transparency and promotion of best practice for countries participating in a forestry mechanism.

Both Annex I and non-Annex I countries are already required to submit National Communications to the UNFCCC, although the content and submission timetables are different in accordance with the principle of common but differentiated responsibilities. Communications contain information on the activities undertaken to implement the Convention. They also set out national circumstances, and provide information on vulnerability, financial resources, technology transfer, and education, training and public awareness. However, only Annex I countries have to provide information on the policies and measures (PAMS) they have put in place to reduce emissions. Annex I countries are also required to submit annual GHG inventories.

National communications from Annex I Parties are subject to an in-depth review conducted by an international team of experts and coordinated by the secretariat. National communications from non-Annex I Parties are not subject to such a review.

The Review recommends that reporting requirements for countries participating in a forestry mechanism should be strengthened. Countries participating in a forestry mechanism should submit annual GHG inventories for the forest sector, and report via national communications on PAMS put in place to reduce deforestation, including an assessment of social and environmental impacts and information on stakeholder consultation processes.

Communications and inventories should be subject to peer review. Transparency in countries' PAMS will open them up to scrutiny from national and international civil society, and from other parties to the Convention, particularly on the impact of their PAMS on the rights and livelihoods of the poor and vulnerable.

Financial transparency
National and local systems and processes that demand accountability and promote transparency in the land-use sector and in new carbon finance streams will be essential. For example, to promote accountability and transparency, Independent Forest Monitors have overseen concession management in Cambodia and Cameroon. A World Bank assessment of the Cameroon programme found that it had enhanced law enforcement, created significant pressure for greater public information and spurred reforms that have increased government revenue collection.[16]

International support mechanisms also have a role. Several countries are applying the Extractive Industries Transparency Initiative (EITI – see Box 12.6) revenue transparency mechanism to forestry. This could be further extended to carbon finance, as a new source of natural resource revenue for governments and companies. A transparent publication system for all forest carbon finance transfers, within and beyond a country's borders, would be a major contribution to ensuring accountability and good financial management. A

---

16 Saunders et al (2008)

national registry for credits would help fulfill this function for international credit transactions, but internal transactions, such as taxes on devolved credit revenue or on how government spends credit revenue, may require additional mechanisms. The Review recommends that forest nations consider establishing a transparency mechanism similar to EITI for this purpose, and that the international community provides appropriate support.

> **Box 12.6: Extractives Industries Transparency Initiative (EITI)**
>
> The EITI[17] aims to strengthen governance by improving transparency and accountability in the extractive industry sector. It sets a global standard for companies to publish what they pay and for governments to disclose what they receive. It supports improved governance in resource-rich countries through the verification and full publication of company payments and government revenues from – principally – oil, gas and mining.
>
> Some countries, such as Ghana and DRC, are already extending their EITI processes to forestry. Expanding EITI or a similar initiative to carbon finance could help to allay fears about the transparent management and distribution of carbon finance revenues.

The freedom and strength of civil society in rainforest nations will be a key factor in determining whether this new resource is managed for the benefit of all of a nation's citizens. Civil society organisations need to develop their understanding of how a new deal on climate change, including forestry, will operate and develop methods, networks and processes for responding to it. Technical and financial support to civil society from the international community can help make this happen.

12.4.2 Voluntary standards and best practice

Programmes to reduce forest loss should avoid unintended harm to the people and places they affect. In addition, voluntary higher standards for programmes that aim to achieve wider benefits such as poverty reduction and the protection of biodiversity could be created. Premium credits generated from such programmes could be given preferential treatment in the international compliance or other markets.

Experience from the CDM suggests that there is little appetite for establishing mandatory sustainability standards within UNFCCC mechanisms.[18] The primacy of national sovereignty in decision-making about land use means that an international agreement on climate change will not be prescriptive in how nations choose to tackle deforestation, beyond the reporting requirements set out above. Nevertheless, the UNFCCC decision on deforestation made at Bali recognised that 'reducing emissions from deforestation and forest degradation in developing countries can promote co-benefits and may complement the aims and objectives of other relevant international conventions and agreements'. Chapter 4 has already described several ways in which policies and measures to reduce deforestation can promote poverty reduction and sustainable development and support other ecosystem services. The Review proposes the development of higher standards, against which compliance can be verified, for those that wish to meet them.

---

17 www.eitransparency.org
18 Hoare et al (2008)

In the voluntary carbon market a range of voluntary standards exists to accredit projects and schemes. Box 12.7 describes some of these in more detail. For a system of national programmes, the Review proposes the following.

1) Development of independent voluntary standards and accreditation processes for assessing national programmes, or parts thereof

A range of standards exists for assessing project-scale carbon forestry projects. Developing criteria for assessing national policies and programmes will be more challenging. It will be crucial that standards, and the institutions developing and assessing them, are trusted by forest nations and potential credit purchasers alike. Criteria for assessing national governance reforms and centrally directed programmes could, for example, relate to the extent of involvement of forest communities in the policy making process, or whether viable livelihoods were developed for forest communities as part of a forest protection programmes. Many forest nations may choose to delegate responsibility for achieving reductions to lower levels of governments, NGOs, companies and individuals. In these cases, standards that operate at sub-national level – probably similar to current project-level standards and potentially instituted by nation states themselves – could be made to apply.

Country sovereignty over what constitutes sustainable development within their borders – and thus what constitutes viable co-benefits – means that a range of different standards is likely to develop, from different centres of influence and perspective. An overarching organisation, such as exists for SFM in the form of PEFC (Programme for Endorsement of Forest Certification Schemes), could help to make sense of a range of standards for credit purchasers. Accreditation processes will also need to be fully transparent to ensure the confidence of both buyers and sellers.

2) Preferential treatment given to premium credits by companies/Annex I countries that are buying credits to meet their commitments

Projects and programmes yielding carbon credits, which include environmental or social co-benefits, are most likely to be developed where investors can see a clear market demand for them. However, achieving demand for co-benefits in the potentially much larger compliance market will require purchasers to give preferential treatment to such credits and to pay a price which reflects the additional costs attached to their generation. There is some potential and precedent for this – for example the UK government's contract for government-purchased offsets requires high and specific standards of environmental integrity and social sustainability. This could be extended to a commitment by sovereign nations purchasing credits to meet their liabilities through purchase of premium credits, or encouraging/obliging those with delegated responsibility (eg through national/regional trading schemes) to do so.[19]

---

19 Hoare et al (2008)

Box 12.7: A selection of voluntary carbon standards

Gold standard

Within the CDM, the best known standard of this type is the Gold Standard. This requires the use of renewable energy and energy efficiency technologies that promote sustainable development for the local community. Gold Standard projects are tested for environmental quality by registered third parties. In 2007, Gold Standard CDM credits were traded at a 15 per cent premium over average annual credit prices, driven in part by Corporate Social Responsibility buyers in the voluntary market.[20]

Climate, Community and Biodiversity Alliance (CCBA)

CCBA has developed voluntary standards to help design and identify land management projects that simultaneously minimise climate change, support sustainable development and conserve biodiversity. The following scorecard shows all 23 criteria, made up of 15 required criteria and 8 optional "point scoring" criteria. To earn CCBA approval, projects must satisfy all 15 required criteria. Exceptional projects that go beyond basic approval may earn a Silver or Gold rating, depending on the number of points scored.[21]

General Section			
G1	Original Conditions at Project Site	Required	
G2	Baseline Projections	Required	
G3	Project Design & Goals	Required	
G4	Management Capacity	Required	
G5	Land Tenure	Required	
General Section			
G6	Legal Status	Required	
G7	Adaptive Management for Sustainability	1 Point	
G8	Knowledge Dissemination	1 Point	
Climate Section			
CL1	Net Positive Climate Impacts	Required	
CL2	Ofsite Climate Impacts ("Leakage")	Required	
CL3	Climate Impact Monitoring	Required	
CL4	Adapting to Climate Change & Climate Variability	1 Point	
CL5	Carbon Benefits Witheld from Regulatory Markets	1 Point	
Community Section			
CM1	Net Positive Community Impacts	Required	
CM2	Ofsite Community Impacts	Required	
CM3	Community Impact Monitoring	Required	
CM4	Capacity Building	1 Point	
CM5	Best Practices in Community Involvement	1 Point	
Biodiversity Section			
B1	Net Positive Biodiversity Impacts	Required	
B2	Ofsite Biodiversity Impacts	Required	
B3	Climate Biodiversity Monitoring	Required	
B4	Native Species Use	1 Point	
B5	Water & Soil Resource Enhancement	1 Point	
Total Project Points			

---

20 Hoare et al (2008)
21 www.ccba.org

> **Plan Vivo**
>
> Plan Vivo generates carbon credits for sale on the voluntary carbon market. Plan Vivo programmes aim to mitigate climate change but also to contribute to poverty reduction. Plan Vivo is a management system used to register and monitor carbon sequestration activities implemented by farmers. Local promoters help farmers to draw up their own work plans, known as Planes Vivos, for forestry or agroforestry systems that reflect their own needs, priorities and capabilities. These are assessed for technical feasibility, social and environmental impact and carbon sequestration potential. Viable plans are registered with the Trust Fund and an agreement for the supply of carbon services via the Fund is signed. The Trust Fund then provides farmers with financial and technical assistance to implement farm- or community-scale forestry and agroforestry developments, on the basis of the carbon that will be sequestered.[22]

12.5 Conclusion

To reduce deforestation effectively and benefit from an international finance mechanism, forest nations will need to make significant reforms to national governance structures and processes. The development of coordinated regional and local mechanisms, each with clear responsibilities, will also be crucial for delivering emission reductions on the ground. Political commitment to the cause of sustainable resource management is the first step to making these changes happen. Inclusive in-country reforms and capacity strengthening are a major opportunity for developing country governments to empower the poor and deliver improved livelihoods for forest communities.

National governments should decide on how best to manage the receipt and use of revenues from forest credits, and this will change over time as new sources of funding come on stream. Where cash transfers to individuals are required, capacity building and demonstration activities to test such approaches will be vital. At the international level, broad-based participation in the governance of carbon revenues would help promote transparency, accountability and strong financial management.

The international community also has a role to play in promoting best practice, policies and measures that deliver reductions in emissions from the forest sector, as well as promoting poverty reduction, biodiversity protection and enhancement. Ensuring that basic safeguards are in place for the people and places who are affected by a new system to reduce deforestation will also be important.

---

22 www.planvivo.org

Part IV: International action, capacity building and short-term funding

The first three Parts of this Review set out the challenge of deforestation; the long-term goal of including the forest sector in an international cap and trade system; and the transitional arrangements and building blocks required over the medium term to achieve this goal. Part IV sets out the action required from the international community in the short term.

The international community must act urgently to address the loss of global forests. In the short term, this will require immediate preparation for linking developing country forests into existing carbon trading schemes; public/private finance to meet the funding gap; and reform of international and national institutions so that forest emissions reductions can be measured, accredited and receive appropriate financial payment.

Chapter 13 sets out the direct action needed to build capacity in the short term. This includes urgent research and analysis; policy and institutional reform; and demonstration activities to test new approaches and demonstrate how credit mechanisms can be used to make production more sustainable, promote afforestation, reforestation and restoration (ARR) as well as securing wider social and environmental benefits. The chapter also discusses the role of public and private finance in meeting the funding gap to finance forest credits in the short to medium term.

Chapter 14 concludes by recognising that climate change and deforestation are global challenges requiring an international response. Action through broad participation and cooperation will be needed as part of the international negotiations under the Bali Action Plan towards a global climate change deal in Copenhagen in 2009. The chapter highlights the need for effective coordination of international financing to support emissions reductions in the forest sector. Finally, it advocates a global sea-change in the way land is used and commodities are produced. This will need to recognise that a shift to more sustainable production methods will be complex and challenging but not impossible if the international community act together effectively.

13. The funding gap and capacity building

Key messages

Strong and urgent action is needed to reduce global deforestation and degradation and promote afforestation, reforestation and restoration. Public and private funding will be required to drive emissions reductions in the short to medium term and there should be access to compliance carbon markets as early as possible.

Many countries will need to undertake a range of preparatory work, reforms and capacity strengthening measures before they can participate fully in a forest carbon credit mechanism.

National solutions to deforestation need to be based on robust analysis of specific national drivers of forest loss and consideration of each country's position on the forest transition curve. This will require research, analysis and knowledge sharing.

Many forest nations will want to undertake policy and institutional reforms, to create a governance environment in which sustainable land and resource management is possible and profitable. Estimates for this Review suggest that necessary reforms and capacity building in 40 forest nations would cost up to $4 billion over five years. Some countries may be able to self-finance, while others may seek overseas development aid support.

Demonstration activities will be needed to test new approaches and demonstrate how credit mechanisms can be used to make production more sustainable; promote reduced emissions from deforestation and degradation (REDD); as well as afforestation reforestation and restoration (ARR); and secure wider social and environmental benefits.

If deforestation is to be halved by 2020, additional public/private finance of $11-19 billion a year to 2020 may be required to fill the funding gap left by compliance carbon markets as they grow.

Funding should be provided by public funds (eg bilateral and multilateral ODA), private investment, and 'pump-priming' of market mechanisms, using a mixture of public and private funds. In the short to medium term, international public funding for forests must be substantially scaled up. Public finance should taper off as carbon markets increase the availability of capital.

International public funds should be coordinated effectively, avoiding a proliferation of competing mechanisms. An overarching secretariat should be established to direct funds and ensure knowledge sharing. Coordination also requires donors and other key stakeholders, such as the UN and the World Bank, to consider climate change and deforestation impacts in their wider assistance programmes with forest nations.

> The design and governance of funds need to be based on equitable participation by developed and developing country governments, and should be done in consultation with indigenous groups and forest communities.

13.1 Introduction

Strong and urgent action is needed to reduce global deforestation and promote afforestation, reforestation and restoration. Public and private funding will be required to drive emissions reductions in the short to medium term and there should be access to compliance carbon markets as early as possible.

Many countries will need to undertake a range of preparatory work, reforms and capacity strengthening measures before they can participate fully in a forest carbon credit mechanism.

The international community needs to act urgently to address global deforestation. In the short term, many developing countries will want support for capacity building to prepare for participation in forest credit schemes. At the same time, a combination of international public and private finance will be needed to meet the funding gap in the short to medium term.

This chapter examines the preparatory work and capacity building that countries will need to undertake in three key areas:

- research, analysis and knowledge sharing;
- policy and institutional reform;
- demonstration activities.

We go on to discuss the short-term funding gap that the international community urgently needs to address. Short- to medium-term public and private finance is essential for investment in major programmes and early crediting mechanisms before and during the period during which national and regional carbon markets will grow and probably merge.

Finally the chapter sets out some key principles of governance of the international architecture for financial support.

13.2 Research, analysis and knowledge sharing

National solutions to deforestation need to be based on robust analysis of specific national drivers of forest loss and consideration of each country's position on the forest transition curve. This will require research, analysis and knowledge sharing.

Chapter 3 set out that the drivers of deforestation are a complex interrelation of global and local factors. Consequently, solutions to the challenge of deforestation need to be based on a sound understanding of the particular drivers of deforestation in specific countries and regions. A first step for forest nations in developing an effective strategy for reducing forest emissions will therefore be robust analysis of national-level rates of deforestation and degradation, along with the country-specific drivers underpinning them.

Forest nations will need to conduct further research and analysis in areas including:

- historical and projected future emissions levels;
- drivers of deforestation;
- capacity for measuring and monitoring;
- capacity of governance, policy and legal frameworks;
- views of stakeholders.

On the basis of this analysis, countries will be able to develop options and ultimately a strategy for:

- developing and agreeing a reference scenario;
- fiscal, institutional and policy reforms;
- implementation channels, legal arrangements and distribution mechanisms;
- social and environmental impact assessment of policies;
- meeting up-front investment for capacity building, equipment, infrastructure, technical assistance and related requirements.

The costs of these kinds of research and analytical work are included in Table 13.1 in the following section. Many countries may seek international technical and financial support to prepare themselves in this way, but the sustainability of solutions to deforestation will depend on the analysis and subsequent strategy development being led by forest nations themselves.

The Forest Carbon Partnership Facility (FCPF) readiness fund and the collaborative programme of UN agencies (UN-REDD) are mechanisms that at the time of writing are being developed to provide channels for support to these activities (see Box 13.6).

As well as knowledge transfer through the provision of technical assistance, the sharing of information and wide dissemination of lessons and best practice will strengthen global capacity for tackling deforestation, and promote the faster development of effective programmes. These should be key functions in an international financing system. The following sections of this chapter examine how this function could operate horizontally across different mechanisms. Forests also cross borders. Consequently, regional and international cooperation on analysis, policy development and information sharing will contribute further to the effectiveness of the global effort. Box 13.1 describes cooperation on information, analysis and capacity building between forest nations through the Poverty and Environment Network.

Box 13.1: Cooperation on research, analysis and information sharing

Poverty and Environment Network – CIFOR

The Poverty and Environment Network (PEN) was launched in September 2004 by the Center for International Forestry Research (CIFOR). PEN is the tropics-wide collection of uniform socio-economic and environmental data at household and village levels from about 30 PEN partners (mainly PhD students), generating a global database with some 5,000 to 6,000 households and 200 to 250 villages from more than 20 countries. PEN research will serve as the basis for the first global comparative and quantitative analysis of the role of tropical forests in poverty alleviation. The data collection includes

> a careful recording of all forest and environmental uses, and income data is collected through four quarterly surveys to shorten recall periods and increase accuracy.
>
> The PEN format represents an innovative way of doing research, involving a large number of partners to collect global data using comparable definitions, questionnaires and methods. PEN will also help in strengthening research capacity in developing countries.
>
> PEN is a six-year project (2004-2010). It is coordinated by CIFOR, but is working closely with resource persons in a number of universities and research institutes on all continents. A grant from the UK is supporting the 2007-2010 phase of data analysis, synthesis and dissemination of results.[1]

13.3 Policy and institutional reform

Many forest nations will want to undertake policy and institutional reforms to create a governance environment in which sustainable land and resource management is possible and profitable. Estimates for the Review suggest that necessary reforms and capacity building in 40 forest nations would cost up to $4 billion over five years. Some countries may be able to self-finance, while others may seek overseas development aid support.

The strategies adopted by countries for policy and institutional reform will differ depending on the drivers of deforestation; current land use and ownership patterns; and political preferences. However, in all cases there are likely to be two key elements:

- Policy and institutional reform: addressing issues such as governance, tenure, land-use planning, tax and other policy drivers.
- Specific activities: reducing deforestation through a range of measures such as tackling illegal logging, sustainable forest management, alternative livelihoods, and protected areas.

The Review commissioned a consortium of organisations to examine the first of these – policy and institutional reform – to assess what common measures most forest countries will need to put in place, and how much this would cost. Box 13.2 describes the work in more detail.[2]

---

1   http://www.cifor.cgiar.org/pen/_ref/home/index.htm
2   Hoare et al (2008)

Box 13.2: Capacity building for policy and institutional reform

This Review commissioned a consortium of organisations – Chatham House, ProForest and the Overseas Development Institute (ODI) – to examine the capacity building requirements for policy and institutional reform in forest nations.

The analysis first undertook an overview of forest and wider land use governance in 25 forest nations, selected to give a cross-section of country and forest circumstances.[3] Issues covered included the state of forest law, land tenure, strength of civil society, the extent of sustainable forest management, sovereign credit ratings and scores on the World Bank Worldwide Governance Indicators index.

Secondly, the consortium examined the minimum governance requirements that countries would need to meet in order to:

- access an international finance mechanism;
- benefit from a mechanism by succeeding in reducing deforestation
- achieve wider benefits such as poverty reduction and biodiversity protection through their programmes to reduce deforestation.

The consortium then estimated costs for the policy and institutional reforms that these 25 countries would need to undertake to meet these governance requirements. The circumstances of some of the 25 were considered directly and some used to cross-check the approach. The figure for 25 countries was then scaled up to 40 countries on the basis that around 40 countries are expected to seek to join a forest mechanism under the UNFCCC.

To provide a global cost estimate for the capacity building needs of rainforest nations, the study considered the types of governance intervention required and then estimated the costs of each of these on the basis of previous programme activities. Defining the methodology was complex because governance requirements are difficult to categorise, and because of the spectrum of functions that might be required to support participation in a future mechanism, ranging from general practices of effective governmental institutions, outside the forest sector but nevertheless essential, to more specific practices that are relevant particularly to the forest sector. There is therefore a wide range of costs that could be included as elements of capacity building for policy and institutional reform.

Once the areas of intervention had been defined, an evidence base was formed from a wide sample of project case studies relating to as many of the interventions as possible. Project data came from a number of donors, including the UK Department for International Development, the World Bank, the International Tropical Timber Organisation (ITTO) and AusAID, as well as from experts who have been involved with implementing and managing projects.

The study estimated a range of costs for each intervention and calculated a range of costs for a generic country by estimating the cost of introducing all of the governance mechanisms. Each of the interventions was assumed to be necessary, to a greater or lesser extent, due to existing gaps in governance capacity. Using this methodology, the consortium estimated the costs at up to $4 billion over five years for 40 forest nations.

---

3   Brazil, Bolivia, Mexico, Costa Rica, Guyana, Venezuela, Colombia, Peru, Indonesia, Malaysia, Papua New Guinea, China, India, Cambodia, Vietnam, Myanmar, Thailand, DR Congo, Congo-Brazzaville, Cameroon, Equatorial Guinea, Liberia, Sierra Leone, Gabon, Ghana.

As set out in Chapter 12, a range of governance factors will need to be addressed. These can be summarised as:

- effective institutions, with clearly defined roles and responsibilities;
- clear and appropriate legislation;
- clear, reliable and equitable land tenure;
- ability to enforce legislation;
- monitoring capabilities.

Some countries may be able to self-finance the necessary reforms and capacity building. Others will seek financial and technical support from external sources. It may be difficult to make any direct link between investment in these types of governance reforms and reductions in emissions, and hence more difficult to fund these activities from carbon revenues. Furthermore, private-sector money is less likely to be invested in countries with poor governance and policy environments. Governance reforms, which create the overarching national context in which on-the-ground action to reduce deforestation can take place, should therefore be an early aspect of implementation of most countries' strategies, before significant carbon revenue funding is available. For all these reasons, public funds will be the most likely source of finance for these activities.

The work commissioned by the Review estimated policy and institutional reform costs for forest nations. Costs for 40 countries are estimated at up to $4 billion over five years (see Table 13.1).

Table 13.1: Cost estimates of readiness for reducing forest emissions in 40 forest nations

Activity	Upper estimate of funding required
Strategy development	$1m
Establishment of relevant infrastructure	$1.5m
Stakeholder consultations	$2m
Pilot testing	$0.5m
Establishment of baseline, monitoring system and inventory	$6m
Land tenure reform	$20m
Land use planning and zoning	$10m
Development of capacity to provide support services for implementation activities, eg RIL, agricultural intensification	$10m
Forest policy and legislation reform	$1m

Activity	Upper estimate of funding required
Tax reform (eg removal of subsidies/tax incentives)	$1m
Standards and guidelines	$1m
Enforcement of planning and environmental requirements, and forest laws	$2m
Independent monitoring	$5m
NGO capacity building	$1m
Effective judicial system	$5m
Institutional reform, clarification of roles and responsibilities, capacity building	$14m
Treasury reform	$5m
Establishment of ability to process and manage payments to project beneficiaries	$5m
5 year costs for one country	$91m
Total 5 year costs for 25 countries	$2.3bn
Total 5 year costs for 40 countries	$3.7bn

Note: Figures are upper estimates

There are two important caveats to be considered when using these figures, which in many cases are derived from past projects carried out mainly with donor funding.

First, costs of previous interventions do not necessarily reflect the actual amounts needed to achieve certain ends – funds spent are more often a reflection of the availability of funds and donor priorities rather than actual requirements. In many cases the costs of a project are a reflection of aid and donor procedures. For example, donors typically have a series of cost levels at which approval can be granted and most projects tend to cluster just under the limit for rapid approval.

Secondly, the projects have not always been successful in achieving the desired outcomes. This could be due to too little being spent, poor project design and management, or to lack of political will.[4]

13.4 Demonstration activities

Demonstration activities will be needed to test new approaches and demonstrate how credit mechanisms can be used to make production more sustainable; promote reduced emissions from deforestation and degradation (REDD) as well as afforestation, reforestation and restoration (ARR); and secure wider social and environmental benefits.

---

4 Hoare et al (2008)

A limited number of programmes have so far specifically aimed to measurably reduce carbon emissions from the forest sector. Of those that exist, the majority have also been at sub-national or project level. Many forest nations will have limited experience of national-level measuring and monitoring techniques; large-scale land use programmes linked to carbon finance; systems for distributing finance; or partnerships between different levels of government, NGOs, the private sector and other stakeholders. Testing approaches for establishing the sustainable production methods described in Chapter 4 through pilot projects or demonstration activities and applying lessons from them will therefore be an important part of early implementation. The Bali decision on addressing deforestation in developing countries encouraged demonstration activities to test new approaches and provided indicative guidance for projects.

Demonstration activities will also play an important role in testing ways in which sustainable production and forest protection measures can promote improved livelihoods and enhance other ecosystem services. At all levels, demonstration activities should aim to strengthen the national and local governance systems within which they are embedded. For example, information transparency mechanisms can be developed at the same time as building measuring and monitoring capacity, or establishing payment for ecosystem services (PES) schemes can include strengthening local governance structures and financial transfer mechanisms.

Funding for demonstration activities can come from a range of sources: public funding (national or international); voluntary carbon markets; and the private sector. Forest nations, potential credit purchasing countries and companies, and private-sector investors should use demonstration activities and their results to learn lessons and improve confidence in a new asset class of forest credits.

Box 13.3 describes some demonstration activities currently underway and planned. Collating and disseminating lessons gathered from demonstration activities will be critical to ensure that lessons are learned and applied in different contexts.

13.3: Demonstration activities currently underway and planned

GMES REDD pilot project: Bolivia and Cameroon

The Global Monitoring for Environment and Security (GMES) initiative, a joint initiative of the European Space Agency and European Union, is working with Bolivia and Cameroon to test the establishment of scientifically valid reference scenarios/baselines for deforestation and, where possible, forest degradation. This will use Earth Observation (EO) technologies and other quantification methods (including field measurements). It also aims to estimate potential future emissions reductions by assessing the carbon dynamics of various forest management strategies (eg traditional logging practice versus reduced-impact logging). The five main components of its work are:

- needs assessments;
- using EO to obtain data on deforestation rates and spatial information on deforestation over a historical period;
- modelling biomass accounting;
- policy scenario analysis;
- technology transfer through capacity building.[5]

5 GAF-AG (2008)

Ulu Masen: Aceh, Indonesia
This project, led by the government of Aceh with the support of NGOs and private project developers, and underwritten by Merrill Lynch, aims to test carbon finance mechanisms to reduce greenhouse gas emissions, contribute to sustainable development and conserve biodiversity.[6] The 30-year project will use land-use planning and reclassification, increased monitoring and law enforcement, reforestation, restoration and sustainable community logging on 750,000 ha of forest in the Ulu Masen Ecosystem and peripheral forest blocks. It is estimated that the proposed activities will reduce deforestation by 85 per cent, leading to 3.4 million tonnes of CO_2 emissions being avoided each year. The project is the first REDD project to have received silver standard certification from the Climate, Community and Biodiversity Alliance (CCBA). The project will sell forward 70 per cent of the generated Verified Emissions Reductions (VERs) with a 30 per cent buffer reserve for leakage and other risks.[7]

Bolsa Floresta: Brazil
Bolsa Floresta rewards communities for their commitment to halt deforestation by distributing payments for ecosystem services to families, communities, and family associations. Families must attend a two-day training programme on environmental awareness and make a zero deforestation commitment. In addition, they must enrol their children in school. They then receive a monthly payment of 50 reais ($30). Community associations can also receive payments of up to 4000 reais ($2500) to support legal income generation activities that do not produce smoke, such as beekeeping for honey production, fish-farming or forest management. Cooperative Investment for administrative support to family associations makes up 10 per cent of the total paid for the families during the year. Bolsa Floresta funds are generated by the interest on a core fund, first established with contributions from the Amazonas government and Bradesco (Brazil's largest private bank).

Deforestation will be monitored on a yearly basis by the Amazonas Sustainability Foundation and the Amazonas State Secretariat for the Environment and Sustainable Development (SDS) team and through satellite images analysed by partner institutions. The programme currently covers six reserves and 2,102 families in six Amazonas State Conservation Units. The objective is to expand to 4,000 families by the end of 2008.[8]

Congo Basin Forest Fund Start-up programme
The Congo Basin Forest Fund, supported by the UK and Norway, will provide near-term funding for a range of new approaches to protect forests and improve livelihoods in the Congo Basin. Start-up funding provided by the UK will finance:

- linking participatory mapping with pilot payments for ecosystem services schemes with communities in the Congo basin;
- institutional strengthening of the Democratic Republic of Congo (DRC) Ministry of Environment;
- piloting community forestry in the DRC through a partnership between the DRC government and NGOs.

---

6   Government of Aceh (2007)
7   Merrill Lynch (2008)
8   http://www.princesrainforestsproject.org/rainforest-nations/the-americas/case-study

13.5 Meeting the funding gap

If deforestation is to be halved by 2020, additional public/private finance of $11-19 billion a year to 2020 may be required to fill the funding gap left by compliance carbon markets as they grow.

Funding should be provided by public funds (eg bilateral and multilateral ODA), private investment, and 'pump-priming' of market mechanisms using a mixture of public and private funds. In the short to medium term, international public funding for forests must be substantially scaled up. Public finance should taper off as carbon markets increase the availability of capital.

If deforestation is to be halved by 2020, additional public/private finance of around $11-19 billion a year to 2020 may be required to fill the funding gap left by carbon markets as they grow (see Chapter 11).

Finance will be required in the short term, before forest credits have access to compliance carbon markets, and in the medium term (to 2020 and beyond), to supplement the carbon market finance which will grow over time. This section focuses on the functions that this short- to medium-term finance is required to fulfil and how it can be delivered.

13.5.1 Public funds

In addition to support for capacity building (discussed in previous sections), public funding will be needed for three main activities in the short to medium term:

- investment in demonstration activities (described in previous sections);
- up-front investment in major programmes (ex-ante finance to deliver emissions reductions);
- 'pump-priming' of market mechanisms, using a mix of public and private funds to reward early action and test crediting mechanisms (ex-post finance for emissions reductions achieved).

Beyond the policy and institutional reforms described in previous sections, reducing deforestation will require forest nations to implement large-scale programmes on the ground, along the lines described in Chapter 4 to establish sustainable production systems and/or alternative livelihoods. Countries may have limited ability to self-finance up-front investments, even with the promise of carbon revenues to reward emissions reductions. This is particularly true of sub-Saharan Africa, where many countries currently have national/sovereign risk profiles that impede private-sector investment in most sectors. Public money (donated, lent or used to underwrite larger private-sector sums) will need to be made available to establish institutions and implement a range of activities if they are to attract private-sector investment or access international carbon markets in the medium term.[9]

Box 13.4 highlights some of the commitments of public funding that have been made already by donors for the type of investments set out above. On current estimates, approximately $4 billion is expected to be made available until 2013. While this might

---

9   Hoare et al (2008)

cover the capacity building costs for preparation, international contributions would need to be significantly scaled up to meet the funding gap and demonstrate real commitment to tackling deforestation. In some countries, absorptive capacity will be a constraint on the support they can accept, particularly before some of the policy and institutional reforms set out in this Review have taken place; however the availability of funds from the international public purse is potentially a major constraint. The diversion of ODA funding for environmental spending is also an issue of concern for many.

> **Box 13.4: Sources of funding to support country programmes for reducing forest emissions**
>
> **United Kingdom**
> The UK's £800 million International Environmental Transformation Fund (ETF) seeks to reduce poverty through environmental protection and helping developing countries respond to climate change. It will support programmes on clean technology and adaptation as well as forests. £50 million of the ETF is supporting the Congo Basin Forest Fund, a multi-donor fund set up to take early action to protect the forests in the Congo Basin region. A further £15 million has been committed to the FCPF.
>
> **Norway**
> Norway is providing $2.5 billion over five years to help reduce carbon emissions from deforestation through its International Climate Change and Forestry mechanism. It is working with the UN and World Bank to ensure appropriate coordination of the international architecture of support to forestry.
>
> **Australia**
> Australia's International Forest Carbon Initiative has so far committed $75 million to a variety of bilateral and multilateral initiatives including capacity building support to Indonesia and a specific partnership on carbon forestry with Kalimantan, a Forest Carbon Partnership with PNG and a research partnership with CIFOR.
>
> **Germany**
> Germany has announced €800 million over four years and a further €500 million a year after 2013 to protect 'forests and other ecosystems under threat'. It has announced a $60 million contribution to the FCPF.

13.5.2 'Pump-priming' market mechanisms, including with private finance

Reductions in deforestation and degradation in developing countries will not be integrated into a carbon market until 2012 at the earliest. However, some forest nations will be in a position earlier to start implementing emissions reductions programmes that can receive international incentive payments. The urgency of addressing deforestation requires that early action is rewarded.

'Pump-priming' covers piloting and scaling up market-type mechanisms for incentivising emissions reduction, in advance of forests being included in a compliance market. This means that forest nations will have early access to national-level crediting systems and incentive payments for reducing deforestation, using the kinds of methodologies (for establishing baselines, measuring and monitoring etc) that a compliance mechanism under the UNFCCC is likely to require. These mechanisms can combine bilateral and multilateral funding with lower-risk investment opportunities for the private sector and will reward early action and build confidence that forest credits can be absorbed into existing carbon markets.

The voluntary market is the only market that can currently undertake projects that generate credits for reduced emissions from deforestation. The size of the Voluntary Carbon Market (VCM – $331 million in 2007)[10] is considerably smaller than the project-based compliance market (CDM value of roughly $8 billion in 2007).[11] The VCM has a much higher proportion of forest credits than the CDM, but projects for reducing deforestation are only 3 per cent of total transactions in the VCM. The VCM also lacks standards to ensure real, measurable and long lasting emissions reductions. The small size of this market, the lower prices on offer and the inability to ensure consistent monitoring and verification standards for emissions reductions indicate that, valuable though the voluntary market is, it will not provide pump-priming for future inclusion of forest credits in a compliance market.

The private sector is currently unwilling to bear the costs associated with the risk of non-delivery of emissions reductions from a national-level government forest programme, or of a new deal on climate change including forestry not materialising. Public funding should therefore be used in the short term to act as a guarantor, effectively bringing down the cost of any liability for the private sector, until the market is large enough to cover these costs. Blending funding from market and non-market sources would help ensure that the price signal to rainforest countries is sufficiently credible to win their participation. Market demand is expected to increase over time, especially if access to the compliance system is in prospect or agreed, so the need for non-market funding sources should diminish. However, a combination of funding is likely to be required for some years into the medium term (see Chapter 11). Figure 13.1 demonstrates how private investment in the forest sector should grow as confidence in the sector increases.

The pump-priming function of pre-compliance credit purchasing facilities helps prepare all actors in a potential future carbon market that includes forest emissions reductions. The benefits for various actors include:

- forest nations: incentives to achieve emissions reductions in a way that generates credits that can be sold on the international market;
- potential private-sector investors: introduces a new asset class of forest credit that they need to prepare themselves for and lowers risks for early investment;
- NGOs: developing standards for generation of forest carbon credits that can later be applied to the compliance carbon market;
- trading schemes, such as the EU ETS, build confidence in legitimacy of a new asset class.

---

10 Hamilton et al (2008)
11 World Bank (2008)

Figure 13.1: Increasing commercial potential of carbon forest finance over time

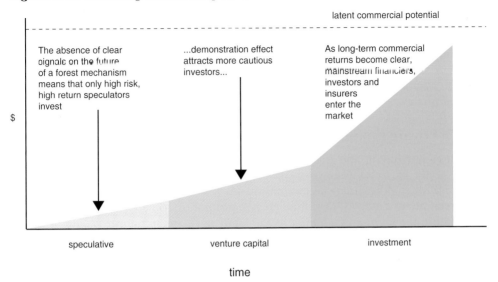

The success of pump-priming will depend on the effectiveness of pre-compliance market credit purchasing facilities. Facilities should:

- reward government-led, large-scale action before a compliance market is in place;
- continue to reward significant-scale sub-national action by countries which are not full participants in a national-level compliance scheme, even when such a scheme is in place;
- reward action in line with the indicative guidelines for demonstration activities from Bali;
- stay in close contact with development of UNFCCC negotiations and adjust requirements gradually and accordingly to ensure credits generated have compliance value if possible;
- guarantee credits for emissions reductions achieved on the basis of initially established criteria, even if these are not later redeemable on the compliance market;
- require high standards with respect to co-benefits such as poverty reduction and biodiversity protection;
- reinvest money from trading credits generated in capacity building in the relevant country;
- work with private-sector investors to ensure that credits generated.

The World Bank's Forest Carbon Partnership Facility (FCPF) Carbon Fund has been developed to fulfil a pump-priming function. The scale of finance currently provided by the FCPF is a small proportion of the finance required, with an initial capitalisation of $300 million. The EU and several individual member states (Italy, the Netherlands, Spain and Denmark) have also established carbon funds managed by the World Bank to purchase credits that can be used to meet their Kyoto obligations and, more broadly, support sustainable low-carbon development in developing countries. These funds could be used to purchase forest carbon credits.

Up-front finance and ex-post rewards can also be linked. Credit purchasing facilities will make payments for emissions reductions once they have been verified. As discussed above, many countries will require up-front funding to finance initial major investments, and there should be communication and coordination between the different funding sources to achieve this. Countries could use carbon revenues to pay back concessional up-front finance.

13.5.3 Private finance

As well as purchasing and trading in forest carbon credits and their prototypes, the private sector may choose to invest in sustainable forest management and ecosystem services through other financial mechanisms. Box 13.5 provides some examples of how traditional finance models (debt financing, securitisation etc) are being rearticulated to work within the forest and land use sector. Many of these mechanisms could be linked to a future carbon market incorporating forestry.

Box 13.5: Examples of private-sector financing for the forest sector

A range of financial instruments can be used to provide funding that is additional to carbon market finance. Examples include equities, bonds, loans and leveraging finance from the insurance industry. Here we examine two examples of innovative financing: the Iwokrama Reserve in Guyana and the concept of ecosecuritisation and forest-backed bonds.

Iwokrama
The Iwokrama Reserve is 370,000 ha of pristine tropical forest in Guyana. In March 2008, Canopy Capital entered into a partnership with the Iwokrama International Centre for Rainforest Conservation and Development (IIC) to measure and then place a value on the ecosystem services (ESS) of Iwokrama's tropical rainforest. Such services include rainfall production, water storage and weather moderation. Canopy Capital is buying a licence to measure and then value the ESS provided by the Iwokrama forest for a period of five years, by making a guaranteed yearly payment to the IIC.
Guaranteed initial income from Canopy Capital will be used by the IIC to continue the sustainable management and conservation of the Iwokrama Reserve and to provide livelihoods for the local communities who depend on the Iwokrama forest. Various approaches to securing substantial investment in ESS are being explored at Iwokrama. In particular, Canopy Capital is looking at marketing ESS through an 'Ecosystem Service Certificate' attached to a ten-year tradable bond, the interest from which will pay for the maintenance of the Iwokrama forest. In the longer term, 90 per cent of any investment upside will go to the IIC for use in this way.[12]

---

12 http://canopycapital.co.uk/resources/Canopy Capital - Frequently Asked Questions.pdf

Forest-backed bonds

A group including Enviromarkets, HSBC, Forum for the Future and the UK's Department for International Development (DFID) have been developing the concept of ecosecuritisation and forest-backed bonds. Through this model, a fund is created to finance investment by sustainable forestry groups. Forestry buyers make a commitment to long-term purchasing contracts for products/services delivered by sustainable forests. In return, forestry and conservation groups make a commitment to long-term product/service supply, sell future payments rights to the EcoSecuritisation Fund and receive immediate payment. Capital market investors buy forest-backed bonds issued in an ecosecuritisation in order to access various tropical forest value streams. One possible structure for the model is set out below.[13]

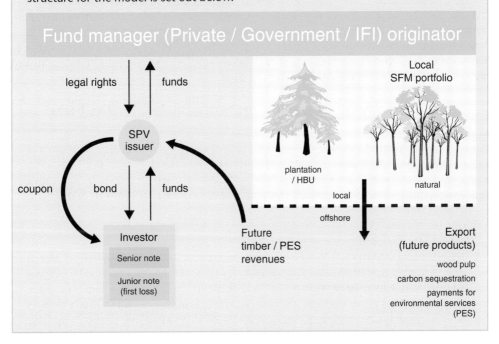

Public funding can also be used to underwrite private finance. The International Finance Corporation (IFC), for example, provides a range of products to support and promote sustainable investment in developing countries through equity finance (buying shares), debt finance (loans) and structured finances such as credit guarantee facilities.

13.5.4 Alternative funding sources and instruments

In addition to bilateral and multilateral public funds, and private finance through market and non-market mechanisms, a range of other funding sources and instruments has been proposed to finance reduced deforestation. Some of the proposals rely on a carbon market function such as levies on carbon market transactions and revenues from the auction of cap and trade allowances. Other funding may come from alternative sources,

---

13 Enviromarkets (2008)

such as international fuel taxation and air travel levies. A selection of these options, and estimates for the amount of funding they could generate are set out in Table 13.2.

Table 13.2: Alternative funding sources for reducing forest emissions[14]

Option	Potential revenue generated
Extend CDM levy to other carbon market transactions • Application of a levy similar to the two per cent share of proceeds from the CDM to international transfers of ERUs, AAUs and RMUs	$10m to $50m Depends on size of carbon markets post-2012
Auction of allowances for international aviation and marine emissions • Sectoral emissions for aviation and marine emissions agreed and auctioned	$10bn to $25bn (aviation) $10bn to $15bn (marine)
International air travel levy • per-passenger charge on international and/or domestic flights	$10bn to $15bn
Funds to invest foreign exchange reserves • Investment of a small proportion of forex reserve holdings into programmes to reduce deforestation	Fund of up to $200bn
Access to renewables programmes in developed countries • Non-Annex I countries eligible for a proportion of money available through Annex I renewable energy programmes	$500m
Debt-for-nature swap • Cancellation of eligible debt in return for agreed investment in reducing deforestation	Further research needed
Tobin tax • A tax of 0.01 per cent on wholesale currency transactions to raise revenue for UNFCCC purposes	$15bn to $20bn
Donated special drawing rights • IMF issues SDRs, Annex I countries donate their allocation to non-Annex I countries	$18bn initially
International Finance Facility • Raise up-front finance from bond markets	Further research needed

---

14 Adapted from UNFCCC (2007). The options here are not exhaustive. Other options include Hare and Macey (2007).

Given the challenges in scaling up public and private funding, these options should be seriously investigated, although there are questions over the feasibility and desirability of all of them. If they can be overcome, forestry will be one of several competing demands for these funds. There is also a need to provide adequate funding for R&D into new clean technologies, technology transfer and climate change adaptation. Consequently, linking the forest sector to national and regional carbon markets as soon as the necessary design elements are in place will be important. Furthermore, decisions over forest funding to top up carbon-market finance will need to be made within a wider context of climate change finance.

13.6 Coordination and governance of public funding

International public funds should be coordinated effectively, avoiding a proliferation of competing mechanisms. An overarching secretariat should be established to direct funds and ensure knowledge sharing. Coordination also requires donors and other key stakeholders, such as the UN and the World Bank, to consider climate change and deforestation impacts in their wider assistance programmes with forest nations.

The design and governance of funds need to be based on equitable participation by developed and developing country governments, and should be done in consultation with indigenous groups and forest communities.

Several donor countries have already stated their intention to support significant investment in sustainable forest and wider land use. A coordinated approach whereby bilateral funds are pooled within multilateral mechanisms and distributed according to agreed criteria for participation and prioritisation should provide a more effective, efficient and equitable delivery of global support to the forest and land use sector (see Figure 13.2).

Figure 13.2: Coordination of funding sources and delivery mechanisms to finance forest emissions reductions more effectively, efficiently and equitably

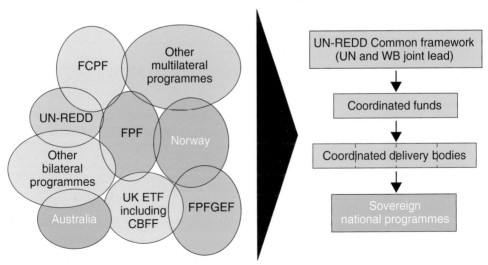

Box 13.6 describes the funding mechanisms currently in place and planned to reduce carbon emissions from the forest sector in developing countries. Both the UN and the World Bank will have a role in overarching coordination and delivery of this support. Several different funds exist or are planned, and there is already potential for overlap and duplication, particularly for supporting 'readiness' functions. A range of mechanisms to suit the needs and preferences of recipient countries will be important, but a proliferation of competing mechanisms should be avoided. The World Bank and the UN in particular need to work together closely to deliver coordinated support and ensure that the division of labour is based on comparative advantage. Figure 13.3 sets out the roles that different funding mechanisms might play. A light-touch overarching secretariat could be established to direct coordination and ensure knowledge sharing. Procedural requirements should as far as possible be in common, in line with Paris Declaration principles on aid effectiveness. So, for example, requirements relating to social impacts of proposals and consultation requirements should be the same across mechanisms.

Governance of funds should be based on equitable participation by developed and developing countries and there should be early consultation on design of mechanisms with forest communities and indigenous peoples. Although the World Bank and the UN are likely to be the overarching holders of funds from bilateral and multilateral sources, mechanisms should be flexible to allow recipient countries to choose their own delivery partners, for example regional development banks, NGOs, bilateral implementing agencies or private-sector organisations. A country-led approach will be essential.

Box 13.6: Multilateral funds to support country programmes to reduce emissions from deforestation

Global Environment Facility (GEF)

The GEF was established in 1991 to help developing countries fund projects and programmes that protect the global environment. GEF grants support projects related to biodiversity, climate change, international waters, land degradation, the ozone layer and persistent organic pollutants. It has financed forest preservation and sustainable land management projects under its land degradation theme and will continue to have a role in pilots and demonstration activities around forests, although new mechanisms set out below are designed to provide large-scale and dedicated finance to the forest sector.

UN-REDD

UN-REDD is implementing a programme, to be implemented over the 18 months running up to COP 15, aims to help prepare countries to access a REDD mechanism through:

- capacity building needs assessment;
- support to strategy development and capacity for monitoring and measuring;
- development and dissemination of guidelines, methods and tools for REDD;
- testing approaches (eg for data management and distribution mechanisms) and collating and disseminating lessons.

Forest Carbon Partnership Facility (FCPF)

The FCPF was launched by the World Bank during the Bali climate talks in December 2007. It is a multi-stakeholder partnership of developing and industrialised countries, NGOs and international financial institutions. The facility's target capitalisation is at least $300 million ($200 million for the carbon fund and $100 million for the readiness fund). At time of writing, a total of $160 million had been committed from donor countries and NGOs.

There are two separate mechanisms that will make up the FCPF:

1. **Readiness Fund:** designed to enable developing countries to have the capacity to participate in a future system that rewards REDD, by supporting the development of measuring and monitoring systems and REDD strategies.
2. **Carbon Fund:** intended to 'pump-prime' crediting mechanisms for REDD. The Carbon Fund will remunerate selected countries in accordance with negotiated contracts for verifiably reducing emissions below a reference scenario.

The FCPF will have a national approach but will not preclude implementation of sub-national programmes and projects.

World Bank Strategic Climate Fund (SCF)

The SCF and the Clean Technology Fund (CTF) together make up the World Bank's new Climate Investment Funds (CIF), a source of interim funding through which the Multilateral Development Banks (MDBs) will provide additional grants and concessional financing to developing countries to tackle climate change.

The SCF will provide financing to pilot new development approaches or to scale up activities aimed at a specific climate change challenge through targeted programmes. A Forest Investment Programme (FIP) of investments to reduce emissions from deforestation and forest degradation through sustainable forest management is currently being developed in conjunction with major donors and developing countries.

Figure 13.3: Respective roles of proposed funding mechanisms for sustainable management of global forests

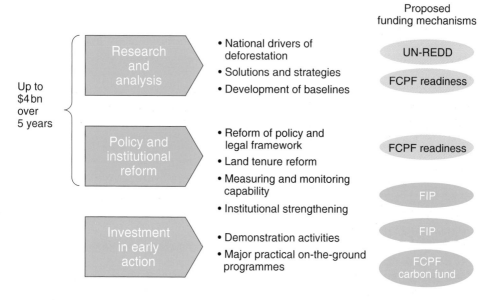

13.6.1 Mainstreaming of climate change and deforestation into donor programmes

Just as forest nations will need to ensure their overall development strategy and policies are consistent with the objective of reducing deforestation, donors will also need to mainstream climate change and deforestation into their development programmes. In this way, programmes that promote deforestation, either directly or indirectly, are avoided. This is likely to require a significant mainstreaming and capacity building effort throughout donor agencies (both bilateral and multilateral). Contributors to major multilateral organisations also have a responsibility to ensure that their contributions are spent in a way that is consistent with broader global objectives on tackling climate change.

Conclusion

A country-led approach is vital to reduce deforestation. However, international support, both technical and financial, will be sought by many forest nations so that they can achieve their national objectives to reduce deforestation and benefit from a new international mechanism on forestry. Considerable work is already underway by bilateral and multilateral organisations to put the finance and mechanisms in place to provide this support. However, there is a heavy responsibility on donor countries, and other key stakeholders such as the UN and the World Bank, to ensure that support is at the scale required and is delivered efficiently and effectively in a way that does not conflict with other global development and environmental priorities.

14. Conclusions

Key messages

Deforestation is progressing rapidly, particularly in the tropics. Firm and urgent action is needed. Otherwise, it is highly unlikely that the international community can achieve a greenhouse gas stabilisation target that avoids the worst effects of climate change.

Action on deforestation needs to be taken as part of the international negotiations under the Bali Action Plan towards a global climate change deal in Copenhagen, and in the wider context of development, poverty reduction and preservation of ecosystem services. The global climate change deal should fully include the forest sector and should set out the arrangements for linking forest credits to regional and national carbon markets.

International action is urgently required to support forest nations in building capacity and preparing for forest carbon finance. Substantial capacity building will be needed in three key areas: research, analysis and knowledge sharing; policy and institutional reform; and demonstration activities.

Climate change and deforestation are global challenges requiring an international response. This will mean action through broad participation and cooperation.

A global step change is needed in the way land is used and commodities are produced. A shift to more sustainable production will be complex and difficult, but not impossible if the international community acts together effectively.

14.1 Introduction

Deforestation is progressing rapidly, particularly in the tropics. Firm and urgent action is needed. Otherwise, it is highly unlikely that the international community can achieve a greenhouse gas stabilisation target that avoids the worst effects of climate change.

The international community needs to act strongly and urgently to address the global loss of forests. Climate change is a major global threat. To avoid the worst effects of climate change, levels of atmospheric CO_2e should be stabilised at 445-490ppm or less. Forestry, as defined by the IPCC, is the third largest source of greenhouse gas emissions, with deforestation in tropical countries the main contributor. If the international community does not tackle deforestation, it is highly unlikely that a CO_2e stabilisation target that avoids the worst effects of climate change can be achieved. Including forests in mitigation efforts can also lower the global costs of meeting an ambitious stabilisation target.

The transition path towards a long-term goal of global cap and trade will need to meet the development needs of countries at different levels of development, particularly the poorest. National sovereignty in this process needs to be respected. The most effective transition path to global cap and trade is likely to be a national, incentive-based approach with increasing finance from emissions trading schemes, combined with additional finance from other sources as carbon markets grow over time.

In the short term, the main objectives should be capacity building and filling the funding gap. Over the medium term, four building blocks are key: effective targets; robust measuring and monitoring of forest emissions; a well designed system for linking forest credits to carbon markets and other sources of finance; and strong governance. In the long term, the goal should be a comprehensive global cap and trade system that fully incorporates the forest sector.

Action on deforestation needs to be part of the international negotiations under the Bali Action Plan towards a global climate change deal in Copenhagen, as well being part of the wider context of development, poverty reduction and preservation of ecosystem services. The key actions that the international community will need to undertake include:

- fully including the forest sector in a post-2012 deal, with arrangements for linking forest credits to carbon markets;
- international cooperation to support capacity building;
- coordinated international action to deliver finance effectively.

14.2 The forest sector in a global climate change deal

Action on deforestation needs to be taken as part of the international negotiations under the Bali Action Plan towards a global climate change deal in Copenhagen, and in the wider context of development, poverty reduction and preservation of ecosystem services. The global climate change deal should fully include the forest sector and set out the arrangements for linking forest credits to regional and national carbon markets.

14. Conclusions

The Bali Action Plan provides a roadmap for the negotiation of a new regulatory framework for international action on climate change, following the expiry of the first Kyoto commitment period in 2012. The Action Plan sets out key areas to be negotiated with a view to reaching a new global climate change deal in Copenhagen at the end of 2009. One area is the reduction of emissions from deforestation and the international finance required to support it. This Review concludes that the forest sector of developing countries should be included fully in a post-2012 transitional system, which should set out the arrangements for linking forest credits to regional and national carbon markets.

The building blocks of this transition in the short to medium term include effective national-level targets (reference levels or baselines); robust measuring and monitoring of emissions reductions below baselines; a well-designed mechanism for linking forest credits to emissions trading schemes; and strong governance.

National sovereignty combined with incentives that reflect the value of reducing forest loss to the global community will be important in making the policy shifts that are needed. Key areas of governance reform that many forest nations will want to take forward include clarifying and securing land tenure and user rights; strengthening the institutional capacity of national, regional and local institutions; and determining and implementing any changes with the full participation of forest communities. Chapter 12 sets out some of the institutional implications of these policy shifts and the options that developing countries may wish to use in the effective distribution of finance sub-nationally.

As well as strong governance at international, national and regional levels, institutional functions will be needed to provide a framework that delivers the remaining three building blocks in the transition: establishing national-level targets; monitoring; and linking to carbon markets. It is beyond the scope of this Review to provide detailed recommendations on the structure of climate change institutions that may cover not only the forest sector but also many other carbon emitting sectors. Furthermore, the particular institutions involved will depend on the climate change negotiations over the next year. However, a set of institutional functions will need to be considered if forest emissions abatement activities are to be effective (see Figure 14.1).

For emissions targets to be effective, baselines will need to be agreed, set and reviewed at regular intervals through international negotiation. National inventories for the forest sector, including net emissions monitored from deforestation, degradation and ARR, will need to be developed, maintained and reported. International verification procedures will also need to be negotiated. Finally, several institutional functions will need to be considered if forest credits are to have access to emissions trading schemes while maintaining the objectives of price stability and incentives for investment in new technology and technology transfer to developing countries. These include national credit registries; a linking mechanism that can aggregate finance from carbon markets and other sources, provide a credit reserve to insure against emissions reductions being reversed; and perhaps a means of reducing investment risk for forest nations that wish to join the scheme (see Chapter 11). Some of these functions already exist with sound institutions in place. Others will need to be created or reformed.

Figure 14.1: Institutional functions for a transitional forest emissions reduction system

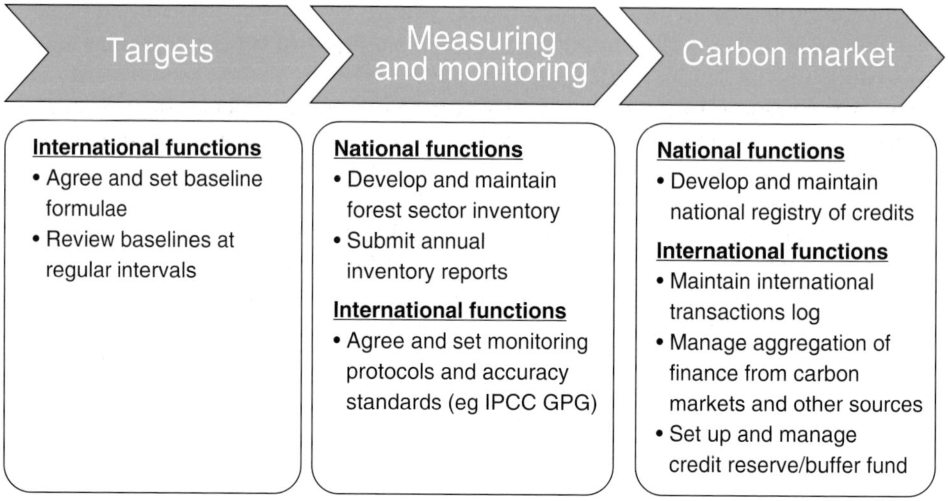

> Targets → Measuring and monitoring → Carbon market

International functions
- Agree and set baseline formulae
- Review baselines at regular intervals

National functions
- Develop and maintain forest sector inventory
- Submit annual inventory reports

International functions
- Agree and set monitoring protocols and accuracy standards (eg IPCC GPG)

National functions
- Develop and maintain national registry of credits

International functions
- Maintain international transactions log
- Manage aggregation of finance from carbon markets and other sources
- Set up and manage credit reserve/buffer fund

14.3 International cooperation to support capacity building

International action is urgently required to support forest nations in building capacity and preparing for forest carbon finance. Substantial capacity building will be needed in three key areas: research, analysis and knowledge sharing; policy and institutional reform; and demonstration activities.

International cooperation is urgently required to support capacity building in forest nations, prior to the introduction of a post-2012 transitional arrangement. In order to benefit fully from the potentially large flows of finance that will be available for reducing forest emissions, many forest nations will need to undertake substantial capacity building in three key areas:

- research, analysis and knowledge sharing;
- policy and institutional reform;
- demonstration activities.

A more accurate picture of the extent of global deforestation and forest degradation is urgently needed. Reductions in forest emissions are only as robust as the data upon which they are based. Many nations have little or no up-to-date data for mapping forest carbon stocks. A first step for forest nations will be robust analysis of national-level rates of deforestation and degradation along with the country-specific drivers underpinning them. Robust measurement of changes in forest cover and land use will need satellite technology, combined with groundwork. An estimated $50 million will be needed for a sample of 25 forest nations to set up national forest inventories, with a further $7-17 million needed for annual running costs.

Many forest nations will want to undertake policy and institutional reforms to create a good governance environment in which sustainable land and resource management is both possible and profitable. Estimates for the Review suggest that necessary reforms and capacity building in 40 forest nations would cost around $4 billion over five years. The strategies adopted by different countries for policy and institutional reform will differ depending on the drivers of deforestation, current land-use and ownership patterns and political preferences. However, they are likely to share common themes such as addressing lack of governance; insecure land tenure and land-use planning; and common measures such as tackling illegal logging, establishing sustainable forest management, promoting alternative livelihoods and extending protected areas.

It will be important to test approaches for establishing sustainable production methods, promoting improved livelihoods and enhancing other ecosystem services through pilot projects or demonstration activities – and apply lessons from them. Demonstration activities should aim to strengthen the national and local governance systems within which they are embedded. Demonstration projects will also be required to test approaches to credit transfers for emissions reductions and to build confidence and ensure that mechanisms and institutions are fit for purpose.

14.4 Coordinated international action to deliver finance effectively

Climate change and deforestation are global challenges requiring an international response. This will mean action through broad participation and cooperation.

If deforestation is to be halved by 2020 and the forest sector to become carbon neutral by 2030, a combination of finance from carbon markets and other funding sources will be needed. Those sources could include various types of private and public sector finance, and the international community will need to agree on the proportion of finance from different sources. Part of the additional funding, including 'pump-priming' of market mechanisms, will need to come from bilateral and multilateral public funds. In the short to medium term, these funds will need to be substantially scaled up before they taper off as carbon markets increase the availability of capital.

If international public funding is to be effective, it will need to be coordinated. Several different funds exist or are planned, and there is already potential for overlap and duplication. A range of mechanisms to suit the needs and preferences of recipient countries will be important, but a proliferation of competing mechanisms should be avoided. To this end, bilateral funds should be pooled within multilateral mechanisms and distributed according to agreed criteria for participation and prioritisation.

An overarching secretariat should be established to direct funds effectively and ensure knowledge sharing. However, the design and governance of funds need to be based on equitable participation by the governments of both developed and developing countries. Although the World Bank and UN are likely to be the overarching holders of funds from bilateral and multilateral sources, mechanisms should be flexible enough to allow forest nations to choose their own delivery partners.

14.5 Conclusion

A global step change is needed in the way land is used and commodities are produced. A shift to more sustainable production will be complex and difficult, but not impossible if the international community acts together effectively.

Climate change is the greatest long-term challenge facing the world. Strong and urgent international collective action is required. To avoid the worst effects of climate change, tackling the loss of global forests must be central to a comprehensive framework for stabilising levels of atmospheric greenhouse gases. With the vast majority of emissions from deforestation now occurring in the tropics, the role of developing countries will be crucial. Any policies to reduce deforestation must be led by sovereign forest nations themselves. In addition, many forest nations will need the support of the international community at large through the provision of expertise and funding for capacity building as well as an international framework that pays forest nations for their global services in reducing forest emissions. Forest nations, international institutions, donors and the private sector all have a role to play.

This Review has examined the options for an international framework that includes the forest sector, and concluded that a global carbon trading scheme is best placed to ensure that emissions from the forest sector are reduced effectively, efficiently and equitably.

The transition path towards a long-term goal of global cap and trade will need to meet the development needs of sovereign nations at different levels of development, particularly the poorest. The most effective transition path is likely to be a national, incentive-based approach with increasing finance from emissions trading schemes combined with additional finance from other sources as carbon markets grow over time.

In the short term, the main objectives should be capacity building and filling the funding gap. Over the medium term, four building blocks are key: effective national-level targets; robust measuring and monitoring of forest emissions; a well-designed system for linking forest credits to carbon markets and other sources of finance; and strong governance. In the long term, the goal should be inclusion of the forest sector within a comprehensive global cap and trade scheme.

The aim of this Review has been to examine and use the available evidence to recommend a practical framework for reducing forest emissions while providing better livelihoods for forest communities and preserving other ecosystem services. In addition, the Review has highlighted the need for more work by scientists, economists and financial and development experts to provide a better knowledge base and analysis of the systems needed to ensure the framework can meet these goals. Areas for urgent research and technical capacity building include more consistent and accurate data on current emissions from the forest sector at national, regional and global levels, and sharing of knowledge and expertise in monitoring emissions, including the use of satellite technology and data management.

The Review has also highlighted the importance of understanding the challenge of deforestation in the wider economic context of agricultural and timber production. Mitigation to reduce deforestation will be successful only if there is a global step change in the way land is used and commodities are produced. This will require a substantial policy shift in three main areas. First, at the international level, we need to place a financial value on forest carbon in a new deal on climate change. Second, at the national level, governance reforms are required to shift policy incentives towards sustainable production. And third, demand-side policies in consumer countries – for example, through preferential procurement of certified products and increasing consumer awareness – can provide incentives for forest nations to promote sustainable production.

This shift to sustainable production will be complex and difficult, but not impossible if the international community acts together effectively. A framework for financing reduced forest emissions will be an essential part of the process, involving the private and public sectors and civil society. Given the consequences of climate change, and the significant contribution of forest emissions towards it, we cannot afford to delay. Together we must act swiftly and decisively.

Bibliography

Angelesen, A (2007) *Forest Cover Change in Space and Time: Combining the vun Thumen and Forest Transition Theories*, World Bank Policy Research Working Paper 4117

Antinori, C and Sathaye, J (2007) *Assessing Transaction Costs of Project-based Greenhouse Gas Emissions Trading*, Lawrence Berkeley National Laboratory, Berkeley, California

Asner, G P et al (2005) 'Selective logging in the Brazilian Amazon', *Science*, 310: 480-482

Attema, E (2005) 'Mission requirements document for the European Radar Observatory Sentinel-1', ESA, ES-RS-ESA-SY-0007, Noordwijk, The Netherlands, http://esamultimedia.esa.int/docs/GMES/GMES_SENT1_MRD_1-4_approved_version.pdf (accessed August 2008)

Australian Government (2008) *Carbon Pollution Reduction Scheme: Green Paper*, Department of Climate Change, Canberra

Ayukai, T (1998) 'Introduction: carbon fixation and storage in mangroves and their relevance to the global climate change – a case study in Hinchinbrook Channel in northeastern Australia', *Mangroves and Salt Marshes* 2: 189–190

Baalman, P and Schlamadinger, B (2008) *Scaling up AFOLU Mitigation Activities in Non-Annex I Countries*, Climate Strategies, London. New work commissioned for the Eliasch Review.

Bala, G et al (2007) 'Combined climate and carbon-cycle effects of large-scale deforestation', *PNAS* vol 104 (16): 6550–6555

Bergeron, Y and Harveu, B (1997) 'Basing silviculture on natural ecosystem dynamics: an approach applied to the southern boreal mixedwood forest of Quebec', *Forest Ecology and Management*, 92: 235-242

Betts, R, Gornall, J, Hughes, J, Kaye, N, McNeall, D and Wiltshire, A (2008) *Forests and Emissions: a Contribution to the Eliasch Review*, Met Office Hadley Centre, Exeter. New work commissioned for the Eliasch Review

Blaser, J and Robledo, C (2007) *Initial Analysis of the Mitigation Potential in the Forestry Sector*, UNFCCC Secretariat, Bern, Switzerland

Bloomberg News, 7 February 2008 *Staples Ends Contracts with Asia Pulp on Environment*, available at http://www.bloomberg.com/apps/news?pid=20601087&sid=acXOtxbg7KDs&refer=home

Bloomgarden, E and Trexler, M (2008) 'Another look at additionality', *Environmental Finance Magazine*, May

Boltz, F et al (2001) 'Financial returns under uncertainty for conventional and reduced impact logging in permanent production forests of the Brazilian Amazon', *Ecological Economics*, Vol 39, pp387-398 (referenced in Chomitz et al (2006))

Bonan, G B (2008) 'Forests and climate change: forcings, feedbacks, and the climate benefits of forests', *Science*, 320: 1444-1449

Braat, L and Ten Brink, P (eds) (2008) *The Cost of Policy Inaction: The Case of Not Meeting the 2010 Biodiversity Target*, Alterra, Wageningen/Brussels

Brack, D (2007) Briefing Paper: *Illegal Logging*, EEDP/LOG BP 07/01, Chatham House, London

Brazilian Government (2004) *Presidência da República – Casa Civil – Grupo Permanente de Trabalho Interministerial para a Redução dos Índices de Desmatamento na Amazônia Legal. 2004. Plano de Ação para a Prevenção e Controle do Desmatamento na Amazônia Legal.* Brasília, DF, Brazil

Brazilian Ministry of Agriculture. *Brazil and Agribusiness Overview.* Brazil

Brown, S et al (2000) 'Issues and challenges for forest-based carbon-offset projects: A case study of the Noel Kempff climate action project in Bolivia', *Mitigation and Adaptation Strategies for Global Change*, 5, 99-121

Brown, S et al (2002) 'Forestry as an entry point for governance reform,' ODI Forestry Briefing No 1, London

Brown, S et al (2008) *Reducing Greenhouse Gas Emissions from Deforestation and Degradation in Developing Countries: A Sourcebook of Methods and Procedures for Monitoring, Measuring and Reporting*, GOFC-GOLD Project Office, Alberta, Canada

Byrd-Hagel Resolution (1997) 105th Congress, 1st Session S RES 98, The National Center for public policy research, Washington DC

Chomitz, K et al (2006) *At Loggerheads: Agricultural Expansion, Poverty Reduction and the Environment,* The World Bank, Washington DC

CIFOR (1998) *CIFOR Annual Report,* CIFOR, Bogor

Colchester, M (1998) *Europe and the World's Forests. Synthesis Report of the European Regional Meeting,* Bonn, 28-29 October

Corbera, E (2007) 'Climate change and forest livelihoods: impacts and synergies', *Arborvitae: the IUCN/WWF Newsletter,* Issue 34, October 2007

DeFries, R S et al (2002) 'Carbon emissions from tropical deforestation and regrowth based on satellite observations for the 1980s and 1990s' in *Procedures of the National Academy of Science US,* 99(22), 1425–1426

Department for Trade and Industry (2007) *Energy Trends June 2007,* A National Statistics publication, London. available at: http://www.berr.gov.uk/files/file40156.pdf

Dutschke, M et al (2005), *Value and Risks of Expiring Carbon Credits from CDM Afforestation and Reforestation,* Hamburgisches Welt-Wirtschafts-Archiv, Hamburg Institute of International Economics, Hamburg, Germany

Ebeling, J and Yasue, M (2008) 'Generating carbon finance through avoided deforestation and its potential to create climatic, conservation and human development benefits', *Philosophical Transactions of the Royal Society for Biological Sciences,* published online, 11 February

EC (2004) *Commission Decision of 29/01/2004 Establishing Guidelines for the Monitoring and Reporting of Greenhouse Gas Emissions Pursuant to Directive 2003/87/EC of the European Parliament and of the Council, C(2004)* 130, European Commission, Brussels, Belgium

Enviromarkets (2007) *Forest-backed Securities: Alternative Finance for Tropical Natural Tropical Forest,* Presentation for West and Central Africa Tropical Investment forum, http://www.itto.or.jp/live/Live_Server/3280/Enviromarket_Grayson.ppt#256,1,Forest-Backed Securities: alternative finance for tropical natural forest

European Environment Agency (2008) *Atmospheric Greenhouse Gas Concentrations (CSI 013) – Assessment,* EEA, Copenhagen

FAO (1990) *Forest Resources Assessment 1990 – Global Synthesis,* FAO Forestry Paper 124, Rome

FAO (2006) Global Ecological Zones http://www.fao.org/geonetwork/srv/en/metadata.show?id=1255&currTab=simple

FAO (2005) *Global Forest Resource Assessment 2005 – Progress towards Sustainable Forest Management,* Forestry Paper 147, Rome

FAO (2008) National Forest Monitoring and Assessment – http://www.fao.org/forestry/nfma.en

Fearnside, P et al (2000) 'Accounting for time in mitigating global warming through land-use change and forestry', *Mitigation and Adaptation Strategies for Global Change*, vol 5, no 3 pp239-270 Springer, Netherlands

Fichtner, W et al (2003) 'The impact of private investor's transaction costs on the cost effectiveness of project-based Kyoto mechanisms', *Climate Policy* 3 pp249-259

Food and Agriculture Organization (2004) *Interactive Wood Energy Statistics*, Wood Energy Programme, Forest Products and Economic Division, FAO, Rome

Food and Agriculture Organization (2006) *Global Forest Resource Assessment 2005: Progress toward Sustainable Forest Management.* FAO Forestry Paper, Rome

Forest Trends (2004) *A New Agenda for Forest Conservation and Poverty Reduction: Making Markets Work for Low-income Producers*, Forest Trends, Washington DC

Franco, M (2008) *Carbon Absorption and Storage*, University of Plymouth, Plymouth. New work commissioned for the Eliasch Review

Frankel, J (2007) 'Formulas for quantitative emission targets' in Aldy, J and Stavins, R (eds) *Architectures for agreement: Addressing Global Climate Change in the Post-Kyoto World*, Cambridge University Press, Cambridge

Gallagher, E (2008) *The Gallagher Review of the Indirect Effects of Biofuels*, Renewable Fuel Agency, East Sussex

Geist, H J and Lambin, E F (2002) 'Proximate causes and underlying forces of tropical deforestation', *BioScience*, Vol 52 (2), pp143-150

GEP (2007) *Global Economic Prospects 2007: Managing the Next Wave of Globalisation.* Available at: http://econ.worldbank.org/WBSITE/EXTERNAL/EXTDEC/EXTDECPROSPECTS/GEPEXT/EXTGEP2007/0,,menuPK:3016160~pagePK:64167702~piPK:64167676~theSitePK:3016125,00.html

Gerbens-Leenes, P W et al (2002) 'A method to determine land requirements relating to food consumption patters', *Agriculture, Ecosystems and Environment*, vol 90, pp47-58

Giambellucat et al (2003) 'Transpiration in a small tropical forest patch', *Agricultural and Forest Meteorology*, 117, 1-22

Gitz, V and Ciais, P (2003) 'Amplifying effects of land-use change on future atmospheric CO_2 levels', *Global Biogeochemical Cycles*, 17

GOFC-GOLD (2008) *Reducing Greenhouse Gas Emissions from Deforestation and Degradation in Developing Countries* http://www.gofc-gold.uni-jena.de/redd/

Government of Aceh (2007) *Reducing Carbon Emissions from Deforestation in the Ulu Masen Ecosystem, Aceh, Indonesia. A Triple-Benefit Project Design Note for CCBA Audit.* Indonesia http://www.climatestandards.org/projects/files/Final_Ulu_Masen_CCBA_project_design_note_Dec29.pdf

Government of Costa Rica (2006) 'Costa Rica: Environmental Services Payments as a policy tool to avoid deforestation and promote forest recovery'. Presentation to UNFCCC REDD workshop, Rome, 30 August – 1 September

Government of Guyana (2008) *Guyana's Low Carbon Development Vision*, Government of Guyana

Grieg-Gran, M (2008) *The Cost of Avoiding Deforestation: Update of the Report prepared for the Stern Review of the Economics of Climate Change*, International Institute for Environment and Development, London. New work commissioned for the Eliasch Review.

Griffiths, T (2007) *Seeing 'RED'? 'Avoided deforestation' and the rights of Indigenous Peoples and local communities*, Forest People's Programme, Moreton-in-Marsh

Gusti, M et al (2008) *Technical Model of the IIASA Model Cluster.* IIASA, Austria. New work commissioned for the Eliasch Review

Hamilton, K. et al (2008) *Forging a Frontier: State of the Voluntary Carbon Market 2008*, Ecosystem Market Place and New Carbon Finance

Hardcastle, P D, Baird, D, and Harden V (2008) 'Capability and cost assessment of the major forest nations to measure and monitor their forest carbon', LTS International, Edinburgh. New work commissioned for the Eliasch Review

Harding, R J and Pomeroy, J W (1996) 'Energy balance of the winter boreal landscape', *Journal of Climate*, 9, 2778-2787

Hare, B and Macey, K (2007) *Tropical Deforestation Emissions Reduction Mechanism (TDERM): A Discussion Paper*, Greenpeace. Amsterdam

Hirsch, A I et al (2004) 'The net carbon flux due to deforestation and forest re-growth in the Brazilian Amazon: analysis using a process-based model', *Global Change Biology*, Vol 10, pp908–924

HM Treasury (2008) Global commodities: a long term vision for stable, secure and sustainable global markets. Available at http://www.hm-treasury.gov.uk/media/7/E/PU579_global_commodities.pdf

Hoare, A et al (2008) *Estimating the Cost of Building Capacity in Rainforest Nations to Allow Them to Participate in a Global REDD Mechanism*, Chatham House, London. New work commissioned for the Eliasch Review

Hoffman, H (2003) 'The Joint Liaison Group between the Rio conventions: An Initiative to encourage cooperation, coordination and synergies', *Work in Progress* Volume 1, Spring 2003, United Nations University

Hooijer, A et al (2006). PEAT-CO_2, *Assessment of CO_2 Emissions from drained peatlands in SE Asia*, Delft Hydraulics report Q3943, prepared in cooperation with Wetlands International and Alterra. Delft Hydraulics, Delft, Netherlands

Hope C (2006) 'The marginal impact of CO_2 from PAGE2002: An integrated assessment model incorporating the IPCC's five reasons for concern', *Integrated Assessment*, 6, 1

Hope, C (2008) *Valuing the Climate Change Impacts of Tropical Deforestation*, Judge Business School, University of Cambridge. New work commissioned for the Eliasch Review.

Hope, C and Castilla-Rubio, J C (2008) *A First Cost-benefit Analysis of Action to Reduce Deforestation*, Judge Business School, University of Cambridge. New work commissioned for the Eliasch Review.

Houghton, R A (1999) 'The annual net flux of carbon to the atmosphere from changes in land use 1850-1990', *The Woods Hole Research Center, Tellus* 51B:298-313

Houghton, R A (2003) 'Revised estimates of the annual net flux of carbon to the atmosphere from changes in land use and land management 1850–2000', *Tellus* B, 55, 2, 378–390

Houghton, R A (2005) 'Tropical deforestation as a source of greenhouse gas emissions' *Tropical Deforestation and Climate Change*, in Moutinho, and Schwartzman, S (2006)

Houghton, R A (2007) 'Balancing the global carbon budget', *Annual Review Earth Planet. Science* Vol 35, pp313–347

House, J I et al (2002) 'Maximum impacts of future reforestation or deforestation on atmospheric CO_2', *Global Change Biology*, 8, 1047-1052.

Huntingford, C et al (2008) 'Towards quantifying uncertainty in predictions of Amazon "dieback"' *Philosophical Transactions of the Royal Society B: Biological Sciences*, 363, 1857-1864

IEA (2007) *World energy Outlook 2007*, OECD/IEA, Paris

IMF World Economic Outlook (2008) *Housing and the business cycle*, available at: http://www.imf.org/external/pubs/ft/weo/2008/01/pdf/text.pdf

IPCC (2000) *Land Use, Land Use Change and Forestry*, Cambridge University Press, Cambridge

IPCC (2001) *Contribution of Working Group I to the Third Assessment Report of the Intergovernmental Panel on Climate*, Cambridge University Press, Cambridge

IPCC (2003) *Good Practice Guidance for Land Use, Land-Use Change and Forestry*, Intergovernmental Panel on Climate Change, IPCC/IGES, Hayama, Japan

IPCC (2006) *IPCC Guidelines for National Greenhouse Gas Inventories, Prepared by the National Greenhouse Gas Inventories Programme*, IPCC/IGES, Hayama, Japan

IPCC (2007) AR4 Synthesis Report, *Summary for Policymakers*, IPPC Fourth Assessment Report, Cambridge University Press, New York

IPCC (2007) WG 1 Chapter 7 'Couplings Between Changes in the Climate System and Biogeochemistry' in *Working Group 1 Report: The Physical Science Basis*, IPPC Fourth Assessment Report, Cambridge University Press, New York

IPCC (2007) WG 3 Chapter 9 'Forestry' in *Working Group III Report: Mitigation of Climate Change*, IPCC Fourth Assessment Report, Cambridge University Press, New York

ITTO (2006) *Status of Tropical Forest Management* 2005, ITTO, Yokohama

Joosten, H and Couwenberg, J (2007) 'Peatlands and carbon' in *Assessment on peatlands, biodiversity and climate change* (eds F Parish, A Sirin and D Charmanet), pp 99-117 Global Environment Centre and Wetlands International

Kaimowitz, D (2002) 'Amazon deforestation revisited', *Latin American Research Review*, vol 37(2), pp221-235

Kanninen, M et al (2007) *Do Trees Grow on Money? The Implications of Deforestation Research for Policies to Promote REDD*, Centre of International Forestry Research (CIFOR), Bogor, Indonesia

Karousakis, K and Corfee-Morlot, J (2007) *Financing Mechanisms to Reduce Emissions from Deforestation: Issues in Design and Implementation*, OECD, Paris, available at: http://www.oecd.org/dataoecd/15/10/39725582.pdf

Kerr, S et al (2004) 'Tropical forest protection, uncertainty, and the environmental integrity of carbon mitigation policies', *Others* No 0411001, EconWPA

Kindermann et al (2008) 'Global cost estimates of reducing carbon emissions through avoided deforestation' in *PNAS*, vol 105, no 30, page 10302, July

Korhonen, L et al (2006) 'Estimation of forest canopy cover: a comparison of field measurement techniques', *Silva Fennica* 40(4): 577–58

Lambin, E F and Geist H J (2003) 'Regional differences in tropical deforestation', *Environment*, vol 45 (6), pp22-36

Landell-Mills, N and Porras, I (2002) *Silver Bullet or Fools' Gold? A Global Review of Markets for Forest Environmental Services and their Impact on the Poor*, IIED, London

Laurance, W et al (1998) 'Tropical forest fragmentation and greenhouse gas emissions', *Forest Ecology and Management* 110: 173-180

Lebedys, A (2004) *Trends and Current Status of the Contribution of the Forestry Sector to National Economics*, FAO, Rome

Lejour, A and Manders, T (1999) 'How carbon proof is Kyoto? Carbon leakage and hot air', *CPB Report* 99/4, 1999, pp43-47

Li, W (2004) 'Degradation and restoration of forest ecosystems in China', *Forest Ecology and Management*, 201, 33-41

Luttrel, C et al (2007) *Forestry Briefing 14: The Implications of Carbon Financing for Pro-poor Community Forestry*, Forest Policy and Environment Programme

Mackenzie, D (1990) '... as Europe's ministers fail to agree on framework for green taxes' *New Scientist*, 29 September 1990. Available at: http://www.newscientist.com/article/mg12717360.900----as-europes-ministers-fail-to-agree-on-framework-forgreen-taxes-.html

Macqueen, D and Vermeulen, S (2006) *Climate Change and Forest Resilience*, Sustainable Development Opinion, International Institute for Environment and Development (IIED), London

Malhi, Y et al (1999) 'The carbon balance of tropical, temperate and boreal forests', *Plant, Cell and Environment*, Vol 22, p 715–740

Malhi, Y et al (2008) 'Climate change, deforestation, and the fate of the Amazon', *Science*, 319, 169-172

Merrill Lynch (2008) *Reducing Carbon Emissions from Deforestation in the UluMasen Ecosystem, Aceh, Indonesia*, Merrill Lynch

Mertens, B and Lambin, E (1999) 'Modelling land cover dynamics: integration of fine-scale land cover data with landscape attributes' *International Journal of Applied Earth Observation and Geoinformation*, 1 (1) 48-52

Michaelowa, A (2005) *CDM: Current Status and Possibilities for Reform*, Hamburgisches WeltWirtshacftsInstitut, Hamburg Institute of International Economics, Hamburg, Germany

Miles, L et al (2008) *Mapping Vulnerability of Tropical Forest to Conversion and Resulting Potential CO_2 Emissions* UNEP World Conservation Monitoring Centre, Cambridge. New work commissioned for the Eliasch Review

Millennium Ecosystem Assessment (2005) *Ecosystems and Human Well-being: Current State and Trends, Volume 1*, Island Press, New York

MINAE (Ministro del Ambiente y Energía) (2008) Decree No. 34371, El Presidente de la República y el Ministro del Ambiente y Energía, Costa Rica

Mitchell, A W, Secoy, K, and Mardas, N (2007) *Forests First in the Fight Against Climate Change: The Vivo Carbon Initiative*, Global Canopy Programme, Oxford

Moat, J et al (2008) *Rapid Forest Inventory and Mapping. Monitoring Forest Cover and Land Use Change* Kew, London. New work commissioned for the Eliasch Review.

Mollicone et al (2007) 'An incentive mechanism for reducing emissions from conversion of intact and non-intact forests'. *Climatic Change*, 83, 477

Molnar, A et al (2004) *Who Conserves the World's Forests?*, Forest Trends, Washington DC

Molnar, A et al (2006) *Community-based Forest Enterprises in Tropical Forest Countries: Status and Potential*, ITTO, RRI and Forest Trends

Monni, S, Syri, S, Pipatti, R and Savolainen, I (2007) 'Extension of EU Emissions Trading Scheme to other sectors and gases: consequences for uncertainty of total tradable amount', *Water Air Soil Pollution: Focus* 7:529–538

Moss, C, Schreckenberg K, Luttrell C and Thassim, L (2005) *Participatory Forest Management and Poverty Reduction: A Review of the Evidence*, ODI, London

Moukinho, P and Schwarkzman, S (2005) (eds) *Tropical Deforestation and Climate Change*, IPAM. Washington DC

Nakicenovic, N and Swart, R (eds) (2000) *IPCC Special Report on Emission Scenarios*, Cambridge University Press, Cambridge

NASA (2008) MODIS web http://modis.gsfc.nasa.gov/data/ [Accessed August 2008]

Neeff, T et al (2005) 'Tropical forest measurement by interferometric height modeling and P-band radar backscatter', *Forest Science* 51 (6) 585-594

Nepstad et al (1994) 'The role of deep roots in the hydrological and carbon cycles of Amazonian forests and pastures', *Nature*, 372, 666-669

Nepstad et al (1999) 'Large-scale impoverishment of Amazonian forests by logging and fire', *Nature*, 398, 505-508

New Zealand Government (2007) *A Framework for a New Zealand Emissions Trading Scheme: Exclusive Summary*, Ministry for the Environment, Wellington

Nobre, C (2008) 'A scientific and technological revolution for the Brazilian Amazon', *Journal of the Brazilian Chemical Society*, 19(3), Editorial

Nordhaus (2008) *A Question of Balance. Weighing the Options on Global Warming Policies*, Yale University Press, US

Oeko-Institut (2007) *Is the CDM Fulfilling its Environmental and Sustainable Development Objectives? An Evaluation of the CDM and Options for Improvement.* Oeko-Institut, Berlin

OECD-FAO (2008) *Agricultural Outlook* 2008-2017. Available at: http://www.agri-outlook.org/dataoecd/54/15/40715381.pdf

Parson, E A and Fisher-Vanden, K (1997) 'Integrated assessment models of global climate change', *Annual Review of Energy and the Environment*, Vol 22: 589-628

Peres, C A et al (2006) 'Detecting anthropogenic disturbance in tropical forests', *Trees*, 21(5): 227-229

Peskett, L and Harkin, L (2007) *Risk and Responsibility in REDD, Forestry Briefing 15*, Overseas Development Institute, London

Peskett, L et al (2008) *Making REDD Work for the Poor.* DRAFT paper prepared by ODI/IUCN on behalf of the Poverty and Environment Partnership

Phillips, O L et al (1998) 'Changes in the carbon balance of tropical forests: Evidence from long-term plots', *Science*, 282, 439-442

Rautiainen, M, Stenberg, P and Nilson, T (2005) 'Estimating canopy cover in Scots pine stands', *Silva Fennica* 39(1): 137–142

Reid, W V and Miller, K R (1989) *Keeping Options Alive: The Scientific Basis for Conserving Biodiversity*, World Resources Institute, Washington, DC

Roberts, D and Nilsson S (2007) *Convergence of the Fuel, Food and Fiber Markets: A Forest Sector Perspective. Paper for the MegaFlorestais Working Group Meeting in St Petersburg, Russia.* October 2007, CIBC and RRI.

Robledo, C et al (2007) 'Climate change and governance in the forest sector: Summary' *Intercooperation*, Berne, Switzerland

Sajwaj, T, Harley, M and Parker C (2008), *Eliasch Review: Forest Managements Impacts on Ecosystem Services*, AEA, Didcot. New work commissioned for the Eliasch Review.

Santilli, M et al (2005) 'Tropical deforestation and Kyoto Protocol: An editorial essay', *Climatic Change*, 71(3): 267–276 (2005)

Sathaye, J et al (2008) *Updating Carbon Density and Opportunity Cost Parameters in Deforesting Regions in the GCOMAP Model*, International Energy Solution (IES) New work commissioned for the Eliasch Review.

Saunders, J, Ebeling, J, Nussbaum, R (2008) *Reduced Emissions from Deforestation and Forest Degradation: Lessons from a Forest Governance Perspective.* Proforest, Ecosecurities and Chatham House, Oxford.

Scherl, L et al (2004) *Can Protected Areas Contribute to Poverty Reduction? Opportunities and Limitations*, IUCN, Gland, Switzerland

Scherr, S, White, A, Kaimowitz, D (2003) 'A new agenda for forest conservation and poverty reduction: making markets work for low income producers' *Forest Trends* Washington DC. Available at:

Schlamadinger, B et al (2006) 'Will joint implementation LULUCF projects be impossible in practice?' available at: http://www.climatefocus.com/downloads/JI_LULUCF _in practice.pdf

Schneider, L (2007) *Is the CDM Fulfilling Its Environmental and Sustainable Development Objectives? An Evaluation of the CDM and Options for Improvement*, Öko-Institut for Applied Ecology, Berlin

Schwarze, R Niles, J and Olander J (2002) 'Understanding and managing leakage in forest-based greenhouse-gas-mitigation projects' *Philosophical Transactions: Mathematical, Physical and Engineering Sciences*, Vol 360, No 1797 Aug 15, 2002, pp 1685-1703, The Royal Society, London

Sentiono, B (2007) *Debt Settlement of Indonesian Forestry Companies: Assessing the Role of Banking and Financial Policies for Promoting Sustainable Forest Management in Indonesia*. Forests and Governance Programme No 11/2007, Centre of International Forestry Research (CIFOR), Bogor, Indonesia.

Sitch, S et al (2003) 'Evaluation of ecosystem dynamics, plant geography and terrestrial carbon cycling in the LPJ dynamic global vegetation model', *Global Change Biology*, 9, 161-185

Sitch, S et al (2005) 'Impacts of future land cover changes on atmospheric CO_2 and climate', *Global Biogeochemical Cycles*, 19, GB2013

Smith, J and Scherr, S (2003) 'Capturing the value of forest carbon for local livelihoods', *World Development*, Volume 31, Issue 12, December 2003, pp 2143-2160

Souza, C et al (2003) 'Mapping forest degradation in the Eastern Amazon from SPOT 4 through spectral mixture models' *Remote Sensing of Environment*, 87, 494-506.

Steininger, M K (2004) 'Net carbon fluxes from forest clearance and regrowth in the Amazon', *Ecological Applications*, 14, S313-S322

Stern, N (2007) *The Economics of Climate Change: The Stern Review*, Cambridge University Press, Cambridge

Stern, N (2008) 'The economics of climate change', *American Economic Review*, 98(2): 1-37

Stern, N (2008) *Key Elements of a Global Deal on Climate Change* available at: http://www.lse.ac.uk/collections/granthamInstitute/publications/KeyElementsOfAGlobalDeal_30Apr08.pdf

Strassburg, B B N et al (2007) 'Reducing Emissions from Deforestation: the "Expected Emissions" approach' (CSERGE Working Paper, January 2007); paper presented at the 26th Meeting of the Subsidiary Board for Scientific and Technological Advice (SBSTA) of the UNFCCC,Bonn, Germany, 8 May

Strassburg, B B N et al (2008 in press) 'Reducing Emissions from Deforestation – A Combined-Incentives mechanism and empirical simulations', *Global Environmental Change*

Streck, C et al (2008) *Climate Change and Forests: Emerging Policy and Market Opportunities*, Chatham House, London

Strengers, B (2004) 'The land-use projections and resulting emissions in the IPCC SRES scenarios as simulated by the IMAGE 2.2 model' *GeoJournal*, 61, 4, 381-393

Sunderlin, W et al (2008) *From Exclusion to Ownership? Challenges and Opportunities in Advancing Forest Tenure Reform*, Rights and Resources Initiative, Washington DC

Sweeting, A and Clark, A (2000) *Lightening the Lode: A Guide to Responsible Large-scale Mining*, Conservation International, Washington DC

Terrestrial Carbon Group (2008) 'How to include terrestrial carbon in developing nations in the overall climate change solution' http://www.terrestrialcarbon.org

The Climate Impacts Group (2002) *The American Forest Foundation as a Carbon Aggregator: Overseeing a carbon offset cooperative for PNW no-industrial private landowner*. Available at: http://cses.washington.edu/cig/outreach/classes/585/files/ES_whitepaper.pdf

Tomaselli I (2006) *Brief Study on Funding and Finance for Forestry and Forest-based Sector; Report to the United Nations Forum on Forests Secretariat*, United Nations, New York

UN Economic and Social Affairs Department (2004) *World Population to 2300*. Available at: http://www.un.org/esa/population/publications/longrange2/WorldPop2300final.pdf

UN Economic and Social Council (2007) *United Nations Forum on Forests. Report of the Seventh Session*, UNESC Official records, 2007. Supplement No 22

UNDP, UNDESA and World Energy Council (2000) *World Energy Assessment*, UNDP, New York

UNEP-WCMC (2008) Forest Restoration Information Service, http://www.unep-wcmc.org/forest/restoration/fris/default.aspx [accessed Aug 2008]

UNFCCC (2006) *Issues Relating to Reducing Emissions from Deforestation in Developing Countries and Recommendations on any Further Process, Submissions from Parties*, UNFCCC Bonn

United Nations (1992) *Report of the United Nations Conference on Environment and Development Annex 3*, available at: http://www.un.org/documents/ga/conf151/aconf15126-3annex3.htm

United Nations (1992) *United Nations Framework Convention on Climate Change*, available at: http://unfccc.int/resource/docs/convkp/conveng.pdf

United Nations (1993) *United Nations Convention on Biological Diversity*, available at http://www.cbd.int/doc/legal/cbd-un-en.pdf

United Nations (1994) *United Nations Convention to Combat Desertification*, available at http://www.unccd.int/convention/text/convention.php

United Nations (1998) *Kyoto Protocol to the United Nations Framework Convention on Climate Change*, available at: http://unfccc.int/resource/docs/convkp/kpeng.pdf

United Nations (2002) *The Rio Conventions: Synergy for Sustainable Development*, available at: http://www.un.org/events/wssd/exhibit/RioConventions.pdf

UNFCCC (2007) *Investment and Financial Flows to Address Climate Change*. UNFCCC, New York, available at: http://unfccc.int/cooperation_and_support/financial_mechanism/items/4053.php

United Nations FCCC (2007) *Report of the Conference of the Parties on Its Thirteenth Session, Held in Bali from 3 to 15 December 2007 Addendum Part Two: Action Taken by the Conference of the Parties at its Thirteenth Session*

University of Michigan (2006) 'Global deforestation', available at: http://www.globalchange.umich.edu/globalchange2/current/lectures/deforest/deforest.html

van Amstel, A R and Swart, R J (1994) 'Methane and nitrous oxide emissions: an introduction', *Fertilizer Research* 37:213-225

Volpi, G (2008) *Brazilian Case Study for RFA Review of Indirect Effects of Biofuels*, submission to the RFA review on ndirect effects of biofuels, published on RFA website http://www.renewablefuelsagency.org, Renewable Fuels Agency.

WBGU (1998) *Die Anrechnung biolischer Quellen und Senken in Kyoto-Protokoll: Fortschritt oder Rückschlang für den globalen Umweltschutz Sondergutachten*, Bremerhaven, Germany

White, A and Martin, A (2004) *Who Owns the World's Forests?*, Rights and Resources Initiative, Washington DC

Wigley, T (2003) *Modelling Climate Change Under No-policy and Policy Emission Pathways*, OECD, Paris

Wilson, D and Dragusanu, R (2008) 'The Expanding Middle: The Exploding World Middle Class and Falling Global Inequality', *Global Economics Paper No 170*, Goldman Sachs, New York

Woodward, S, Roberts, D L and Betts, R A (2005) 'A simulation of the effect of climate change-induced desertification on mineral dust aerosol', *Geophysical Research Letters*, 32

World Bank (2004) *Sustaining Forests: A Development Strategy*, World Bank, Washington DC

World Bank (2007) *Global Economic Prospects 2007*, World Bank, Washington DC

World Bank (2008) *State and Trends of the Carbon Market 2008*, World Bank, Washington DC

World Bank (2008) *World Development Report 2008: Agriculture and Development*, World Bank, Washington DC

World Bank (2008) *Worldwide Governance Indicators, 1996–2007*, World Bank, Washington DC

WWF (2002) *The World Summit on Sustainable Development*, available at: http://www.wwf.org.uk

Wunder, S (2000) *The Economics of Deforestation: The Example of Ecuador*, Macmillan, St Antony's Series, Houndmills, Basingstoke

Yamin, F and Depledge, J (2004) *The International Climate Change Regime: A Guide to Rules, Institutions and Procedures*, Cambridge University Press, Cambridge

Zbinden, S and Lee, D (2004) 'Paying for Environmental Services: An Analysis of Participation in Costa Rica's PSA Program', *World Development*, Volume 33, Issue 2, February 2005, Pages 255-272

Index

AAUs *see* Assigned Amount Units
abatement
    cap and trade systems 96, 98
    carbon markets linking 165–190
    costs 75–77, 88–89, 94
    low-cost 69, 70, 189
    see also mitigation
above-ground carbon densities 147, 148–149
ABS *see* access and benefit sharing
absorptive capacity 223
access
    carbon 136
    carbon markets 126–127
    emission trading 123, 125, 127
    international finance 122
access and benefit sharing (ABS) 103
accountability 201, 202, 203, 206, 207
accreditation 101, 115–116, 161–162, 208
Aceh, Indonesia 221
adaptation 11, 53, 55
additionality
    baselines 129, 130, 134, 135, 136, 138, 140, 141–142
    long-term framework 83, 86, 87–88, 91, 92
    opportunity costs 74
Ad Hoc Technical Expert Group (AHTEG) 104
administrative level 205
aerial photography 151, 152–153
afforestation 91, 105, 145, 155, 156
afforestation and reforestation (A/R) 101, 106, 109, 110, 115, 130, 186
afforestation, reforestation and restoration (ARR)
    cap and trade systems 83, 96, 97
    carbon cycle impacts 20–22
    carbon markets 168, 183
    carbon stocks 7
    demonstration activities 213, 219
    increasing 33
    long-term framework 127
    natural disturbances 88
    promoting 33, 211
    rates of 23–24
AFOLU *see* agriculture, forestry and other land use
Africa 23, 27, 36–37, 46, 53, 114, 147, 222
Agenda 21 104
aggregation of funding sources 184–185
agreements 61, 102–106
agricultural products 35, 39–40, 49, 63, 66
agriculture 8, 24, 43, 47–48, 53–54, 200, 238

agriculture, forestry and other land use (AFOLU) 159, 160
    see also land use, land-use change and forestry
agroforestry 54, 55, 56
AHTEG *see* Ad Hoc Technical Expert Group
aid 95, 219
air travel levies 228
albedo 25, 26
alternative employment 58–60
alternative funding sources 227–229
Amazon region 10, 27, 36, 59–60, 157–158, 221
amplifying effects 25
Annex I countries
    carbon markets 167, 168, 169, 170, 172, 174, 175, 189
    current international framework 101, 116
    emissions trading 165
    Kyoto Protocol 105, 106, 107, 108, 109, 112, 113
    long-term framework 123, 124
    measuring and monitoring 161
    reporting 206
    UNFCCC 104, 105, 106
    voluntary standards 208
Annex II countries 104
Annual Deforestation Rate Assessment (PRODES) 157, 158
anthropogenic emissions 1, 6, 7, 8, 21, 24, 101, 105, 107, 111
    see also carbon dioxide; greenhouse gases
Arctic sea ice 1, 2
ARR *see* afforestation, reforestation and restoration
Asia 23, 27, 37, 53
Asia Pulp and Paper 67
Assigned Amount Units (AAUs) 109, 170
atmospheric carbon dioxide *see* carbon dioxide
AusAID 217
Australia 170, 171, 173, 223
availability of land 49, 50–51
average baselines 135–136, 136–138
aviation emissions 228
avoided deforestation 72–74, 75, 129, 133, 138, 140, 143
awareness raising 239

Bali Action Plan, 2007 1, 10–11, 117, 131, 138, 211, 220, 233, 234, 235
Bandundu, DR Congo 194
banking sector 204
baselines 124, 129, 130–143, 168, 183, 184, 188, 235
BAU *see* business as usual

benefits 71, 77–80, 85, 89
best practice 146, 205, 207–210, 215
bilateral funding 196, 213, 222, 229, 237
binding caps 107, 124, 186
biodiversity 10, 53, 54, 58, 103, 154, 207
biofuels
　certification 66
　drivers of deforestation 35
　food prices 39, 40
　land 50, 54
　slowing expansion 67
　sustainable production 49, 64
biomass 16, 18, 19, 26
biophysical aspects 25–26
biospheric services 10
BirdLife Indonesia 61
BirdLife International 152
Bolivia 155, 220
Bolsa Floresta, Brazil 199, 221
bonds 226, 227
boreal regions 149
bottom-up approaches 74
boundary demarcation 61
Brazil
　deforestation drivers 41, 44
　measuring and monitoring 157–158
　PES 199, 221
　sustainable production 53, 54, 57, 58, 66–67
　technology centres 59–60
buffers 187
　see also reserves
burning 19, 20
Burung Indonesia 61
bushmeat 9
business as usual (BAU)
　additionality 87
　baselines 130, 133, 134, 136, 138
　carbon markets 182
　costs 75, 77
　deforestation 26–28, 29, 30, 31, 32
　effective reduction targets 129, 130
　emissions 85
　welfare 4

calibrating satellite data 145, 147
Cameroon 41, 152–153, 206, 220
Canada 147, 171, 174
Canopy Capital 226
canopy cover 20, 157, 158
capacity 45–46, 146, 195, 200, 204
capacity building
　costs 70, 71
　governance and finance distribution 197, 210
　institutional 191
　international 211–232, 233, 234, 235, 236–237, 238
　long-term framework 121, 125, 126, 238
　measuring and monitoring 122, 126, 127, 145, 159, 161, 162–163
　sustainable production 54, 64
　technological 59
caps 12, 105, 107, 108, 113, 116, 124, 175, 186
cap and trade systems
　carbon markets 167, 186, 190
　current framework 102
　incentive-based 124–125
　international emissions reduction 90
　leakage 87
　long-term framework 92–98, 99, 119, 121, 122, 123, 234
　mitigation costs 75, 81, 83
　sustainable production 64
　US 171
carbon
　cycle 15, 16–18, 18–22
　dependency 62
　emissions 49, 52, 83–99
　finance 54, 55, 99
　neutrality 96, 97, 237
　reserves 187–188
　stocks 55–56, 107, 145, 147–155, 162
　taxation 91–92
　trading 75–77, 81, 113–114, 238
　value 60, 62, 63–64, 84–85, 90, 238
　see also emissions reduction; sequestration
carbon credits
　capacity building 213, 214, 224, 226
　cap and trade systems 97
　carbon markets 172
　distribution of finance 198
　governance 198, 210
　mitigation costs 72, 75, 76
　supply and demand 127
carbon dioxide
　CERs 110
　deforestation 15, 19, 29
　efficiency 88
　sequestration 24–26
　sink effect 21, 22
　stabilisation 1, 5, 6, 7, 84, 85, 172, 234
　uncertainty 26, 27, 28
　see also emissions reduction; greenhouse gases
Carbon Fund 231
carbon markets
　compliance 213, 214, 222, 224
　current international framework 116
　developing countries 114, 119, 175, 223
　emissions trading 123, 124
　finance 201–202, 227
　forest credits 121, 126, 127, 223, 234, 238
　linking to 126, 127, 165–190, 211, 233, 235
　transitional systems 236
　voluntary standards 210
　see also emissions trading
carbon sinks 16, 21, 25
carbon stocks 6, 7, 15, 55–56
Caribbean 153–154

Index   253

Carnegie Landsat Analysis System 158
cash distribution 200
cash reward payments 196
cash transfers 191, 197, 200–201, 210
CBD *see* Convention on Biological Diversity
CCBA *see* Climate, Community and
    Biodiversity Alliance
CDM *see* Clean Development Mechanism
CENADEP *see* Centre National d'Appui an
    Development et a lá Participation
    Populaire
Center for International Forestry Research
    (CIFOR) 215, 216
Central Africa 46, 163
Central America 23
central carbon reserves 187–188
Centre National d'Appui an Development
    et á la Participation Populaire
    (CENADEP) 194
CEPF *see* Critical Ecosystem Partnership Fund
cereal productivity 53
CERFLOR 66
CERs *see* certified emission reductions
certification 49, 64, 65, 66, 115, 221
Certified Emission Reductions (CERs) 108–109,
    110, 114, 115–116, 186
Chatham House 217
China 21, 23, 62, 65, 112, 114
Chomitz, K 62
CIFOR *see* Center for International
    Forestry Research
Cikel Brasil Verde Madeiras Ltda 66–67
civil society 201, 206, 207, 239
Clean Development Mechanism (CDM)
    alternative funding sources 228
    baselines 132
    carbon markets 180, 186
    current international framework 101,
        105, 106, 108, 109, 110, 111, 113,
        114–115, 117
    effective reduction targets 129, 130
    project-based approaches 87
    standards 207, 209
cleaner technologies 165, 167, 174
Clean Technology Fund (CTF) 231
clear-cut logging 20
Climate, Community and Biodiversity Alliance
    (CCBA) 209, 221
coastal regions 4
cold regions 26
collective action 81, 238
Colombia 45
combined funding sources 224
combined (historical and average) baselines
    136–138, 140, 141, 143
COMIFAC *see* Commission des Forets d'Afrique
    Centrale
command and control measures 197
commercial potential 225

Commission des Forets d'Afrique Centrale
    (COMIFAC) 163
commitments 102, 103–104, 107–109, 112, 117,
    123–124, 210
commodities, demand for 49, 50, 51, 52
commodity production 233, 238
common but differentiated responsibilities 91, 104,
    110, 206
community level
    deforestation 43
    demonstration activities 221
    distribution of finance 199, 205
    ecosystem services 8–10
    environmental stewards 49, 55, 56
    forest management 56–57
    livelihoods 238
    mapping 194
    protected areas 60–61
    public funding 229, 230
    see also indigenous communities; local level
compensation 92, 197
competing mechanisms 213, 229, 230, 237
compliance carbon markets 213, 214, 222, 224
conditionality 61
Conference of the Parties (COP) 104, 110, 117
conflicts 45
Congo Basin 57, 58, 157, 163, 221, 223
conservation 60–62, 69, 70, 74, 95, 106, 152–153, 154
Conservation International 58, 155
consistency 91, 145, 146, 159, 160, 161, 164
consumer countries 62, 64–68, 239
consumption 4
continental forest cover 23, 27
Convention on Biological Diversity (CBD) 103, 104
conventions 99
cooling effects 25
cooperation 146, 162, 163, 211, 215, 233, 234,
    236–237
cooperatives 57
coordination 91–92, 104, 210, 213, 229–230,
    234, 237
COP *see* Conference of the Parties
Copenhagen, 2009 1, 10, 11, 112, 117, 211, 233,
    234, 235
Corporate Social Responsibility 66
Costa Rica 62, 198
costs
    abatement 75–77, 88–89, 94, 169
    capacity building 237
    cap and trade systems 94, 97
    carbon dioxide externalities 85
    carbon markets 174
    climate change impacts 1, 2
    damage 30, 31, 32, 77, 78
    deforestation 13, 28–32, 63, 69
    delegation 191
    distribution of finance 89, 197
    forest emissions 7, 15
    GHGs stabilisation 6, 88, 98

costs (*continued*)
  loss of ecosystems 36, 39, 63–64
  measuring and monitoring 113, 119, 146, 158, 162–163
  mitigation 13, 69–80, 95, 142
  policy and institutional reform 213, 216, 217, 218–219
  remote sensing 151
  research and analysis 215
  sampling 155
  satellite imagery 150
  transaction 72, 88, 101, 115, 117, 196, 200
  voluntary standards 208
country-specific approaches 53, 192, 232, 236
cover, forest 15, 17, 23, 48
credibility 122
credits
  avoided deforestation 133
  baseline 129, 130–143, 183, 184, 187
  carbon markets 165, 166, 167, 182, 187–188
  distribution 200
  international 174, 175–177, 179, 180–181, 189
  trading 109–110
  transfers 122
  see also carbon credits; forest credits
Critical Ecosystem Partnership Fund (CEPF) 155
cross-government policy 53
CTF *see* Clean Technology Fund
current international framework 101–118
customary rights 193

damage costs 30, 31, 32, 77, 78
Darwin Initiative 154
databases 215–216
data interpretation techniques 151
deadwood 19, 158
debt-for-nature swap 228
decision-making 45–46
decomposition 19
deforestation
  avoided 72–74, 75, 129, 133, 138, 140, 143
  Bali Action Plan 117
  baselines 131, 133, 135, 136, 137–138, 139–141
  cap and trade systems 96
  challenge of 13–80
  current frameworks 122
  demonstration activities 221
  developing countries 36, 48, 223, 238
  emissions reduction 6–7, 129
  funding 196
  GHG emissions 8
  governance 192, 195, 210
  Kyoto Protocol 113
  mainstreaming 232
  measuring and monitoring 145, 155, 156, 157–158, 160
  mitigation costs 95
  multilateral funds 230–231

  national 113, 131, 236
  policy safeguards 205
  research and analysis 214–215
  short to medium term 123
  urgent action 213, 214, 223, 233, 234
  voluntary standards 207
degradation
  carbon cycle 19–20, 33
  emissions reduction 6–7
  Kyoto Protocol 113
  measuring and monitoring 145, 155, 158–159, 160
  national 236
  rates of 24
  research and analysis 214–215
  see also reducing emissions from deforestation and degradation
delaying emissions 88
delegation 191, 192, 197–201, 208
demand
  agricultural products and timber 39–40, 49
  carbon markets 168–174
  commodities 49, 50, 51, 52, 68
  emissions trading 167
  international credits 165
  middle class growth 37
demand-side measures 49, 50, 52, 62, 64–68, 239
Democratic Republic of Congo 44, 58, 194
  see also Congo Basin
demonstration activities
  governance and finance distribution 191, 195, 197, 204, 210
  international action and capacity building 121, 126, 211, 213, 214, 219–221, 225, 236, 237
Department for International Development (DFID), UK 217, 227
dependency 9, 52–53, 62
desertification 103
Designated National Authority (DNA) 106
DETER *see* Near Real Time Deforestation Detection
DETEX *see* Forest Exploitation Detection System
developed countries
  alternative funding sources 228
  baselines 141–142
  cap and trade systems 94, 98, 122
  current international framework 101
  deforestation 42, 48
  distribution of finance 237
  measuring and monitoring 159
  responsibility 92, 104
  UNFCCC and Kyoto Protocol 106
  UN Rio Conventions 103
developing countries
  Bali Action Plan 117
  baselines 136
  capacity building 214
  cap and trade systems 94, 98, 122
  carbon dioxide emissions 24

Index 255

carbon markets 114, 119, 165, 168, 174, 175, 223
  current international framework 102
  deforestation 36, 48, 223, 238
  development stage 119, 122
  distribution of finance 196, 237
  emissions reduction 7, 89
  forest sector inclusion 235
  Kyoto Protocol 101, 108, 112, 113
  measuring and monitoring 159, 161
  middle class 37
  mitigation costs 80
  population 35, 37
  public funding 230, 231
  transition pathways 124–125
  UNFCCC and Kyoto Protocol 106
  vulnerability of 2
  see also non-Annex I countries
development 43, 119, 121, 122, 193, 200, 232, 234
devolution 45, 197–201
DFID *see* Department for International Development
diet 37
differentiated national emissions 94
differentiated responsibilities 94, 104, 110, 206
discounting 30
distribution
  cap and trade systems 94
  carbon stocks 147
  CDM 114
  costs and benefits 89
  finance 117, 119, 191–192, 196–204, 205, 237
disturbed forests 19, 147
DNA *see* Designated National Authority
domestic carbon taxation 91
donated special drawing rights 228
donors 213, 219, 222–223, 229, 232
double counting 199
drivers of deforestation 13, 35–48, 214
droughts 2
dust aerosol emissions 26

early action 232
earmarked funds 202, 203
Earth Observation (EO) 220
Earth Remote Sensing Satellites 151
East Asia 53
ecological services 10
ecological zones 147, 148
economic level 1, 2, 15–33, 35–48, 93, 238
ecosecuritisation 226, 227
ecosystems 4
ecosystem services
  adaptation 53
  carbon stocks 55–56
  damage costs 78
  deforestation 30, 36, 39
  demonstration activities 220
  emissions reduction 7

forests 1, 8–10, 33
  investment 226
  payment for 49, 61–62, 72, 196, 198, 199, 220, 221
  preserving 238
  sustainable production 49, 63–64
education 39
EE *see* expected emissions
effectiveness 46, 81, 83, 85, 86–88, 94, 132–133, 139–140, 200
effective reduction targets 129, 130–143
efficiency
  baselines 132–133
  cap and trade systems 81, 92, 94, 95, 97
  carbon markets 175
  carbon taxation 91
  emissions trading 113
  long-term framework 83, 85, 88–89
  sustainable production 49, 50, 52, 62
EITI *see* Extractive Industries Transparency Initiative
EMF *see* Energy Modelling Forum
emission reduction units (ERUs) 108
emissions
  carbon dioxide 24–26
  measuring and monitoring 145–164
  reversal of 184, 186–188, 235
  rights 109
  see also carbon dioxide; greenhouse gases
emissions reduction
  Bali Action Plan 11
  carbon markets 166–190
  current international framework 101–118
  effective targets 129–143
  governance and finance distribution 191–210
  inclusion of forests 7
  international action 211–239
  long-term framework 12, 83–99, 121–127
  mitigation costs 69–80
  sustainable production 52
  see also reducing emissions from deforestation and degradation
Emissions Reduction Units (ERUs) 110
emissions trading
  carbon markets 168, 169–182
  current framework 101, 106, 108–111, 113–114
  long-term framework 119, 121, 122, 123, 125, 127, 234, 235, 238
  see also carbon markets
Emissions Trading Scheme (ETS), EU 108, 165–166, 167, 170, 171, 173, 175–177, 177–179, 184, 189
employment 9–10, 49, 58–60, 62
energy 8, 25, 186
energy efficiency 6, 209
Energy Modelling Forum (EMF) 172
enforcement 35, 45, 65, 197, 201
Enviromarkets 227

Index

environmental impact assessments 58, 191, 192, 205
environmental services (ES) 61–62, 200
environmental stewardship 49, 55, 56, 200
EO *see* Earth Observation
equity
  baselines 132–133, 135, 139–140
  cap and trade systems 81, 94
  carbon taxation 92
  distribution of finance 214, 237
  long-term framework 83, 85, 89, 98
  PES 62
  public funding 229, 230
ERUs *see* emission reduction units
ES *see* environmental service
ETF *see* International Environmental Transformation Fund
ETS, European Union (EU) 108, 165–166, 170, 171, 173, 175–177, 177–179, 184, 189
EUA *see* EU Allowance
EU Allowances (EUAs) 175, 176, 177, 178
Europe 23, 42
European Space Agency 220
European Union (EU)
  demonstration activities 220
  pump-priming 225–226
  regulations 65, 90
  sustainability certification 66
  see also Emissions Trading Scheme
evaporation 25, 26
evapotranspiration 25
exclusion 96–97
Executive Board, CDM 111
expected emissions (EE) 138
expert knowledge 152, 162–163, 238
exports 39
ex-post rewards 226
extensification 54
externalities 63–64, 84, 85, 90, 91
Extractive Industries Transparency Initiative (EITI) 206–207
extreme weather 39

FAO *see* Food and Agriculture Organisation
farmland 19
FCPF *see* Forest Carbon Partnership Facility
FIP *see* Forest Investment Programme
fires 147
fixed prices 185
FLEGT *see* Forest Law Enforcement, Governance and Trade
flexibility 92, 94, 113–115
FONAFIFO 62, 198
food 4, 9, 38, 39, 40
Food and Agriculture Organisation (FAO) 24, 51, 147, 148, 163
foreign exchange reserves 228
forest-backed bonds 226, 227
Forest Carbon Partnership Facility (FCPF), World Bank 163, 215, 225, 231, 232

forest credits
  carbon markets 12, 166–190, 222, 233, 234, 235, 238
  mitigation costs 70, 71
Forest Exploitation Detection System (DETEX) 158
Forest Fund Start-up programme, Congo Basin 221
Forest Investment Programme (FIP) 231, 232
Forest Law Enforcement, Governance and Trade (FLEGT) 65, 66, 90, 195
Forest Stewardship Council (FSC) 66–67
Forum for the Future 227
fossil fuels 8, 21, 22
fragmentation 20, 24
FSC *see* Forest Stewardship Council
fuel taxation 50, 228
fuelwood 9, 68, 157
full survey techniques 153–154
funding 95, 106, 196, 220
funding gap
  carbon markets 166, 182, 183–184
  international action 211, 213–232, 234, 238
  long-term framework 119, 121, 122, 125, 126
future aspects 26–32, 112, 136

G8 Summit, Heiligendamm, 2007 85, 96
Gallagher Review, 2008 50–51, 53, 54
gas prices 86, 88
GCOMAP 75
GEF *see* Global Environment Facility
geology 147
Germany 223
GHGs *see* greenhouse gases
global biome averages 162
global climate change deal 234–236
Global Environment Facility (GEF) 106, 111, 230
Global Monitoring for Environment and Security (GMES) 220
global positioning satellite (GPS) 194
global warming *see* temperature
GLOCAF *see* Office of Climate Change Global Carbon Finance
GMES *see* Global Monitoring for Environment and Security
GOFC-GOLD 161
Gold Standard 209
Good Practice Guidance (GPG), IPCC 145, 146, 159, 160, 164
goods and services 6
governance
  baselines 142
  cap and trade systems 99
  certification 66
  demonstration activities 220
  distribution of finance 202, 204
  drivers of deforestation 36, 46, 48
  international 205–208, 234, 235
  long-term framework 121, 126, 127, 237, 238
  medium-term approach 119, 191, 210
  mitigation costs 70

policy and institutional reform 213, 216, 217
public funding 229–230
sustainable production 62, 64, 68
government
    baselines 131
    deforestation 46, 197
    distribution of finance 200, 201–204
    reform 192–195
    sustainable production 53, 66, 68
    valuing carbon and ecosystem services 64
GPG *see* Good Practice Guidance
GPS *see* global positioning satellite
grants 203
greenhouse gases (GHGs)
    emissions 1, 6, 24
    stabilisation 5–6, 33, 84, 172, 233, 234, 238
    UNFCCC 101, 103
    see also carbon dioxide
groundwork 145, 150, 151–152, 153, 155, 158
growth policies 43
guaranteed prices 77, 188–189
Guyana 193, 200, 226

harmonisation of taxation 91, 92, 94
harvested wood 19
health 4
heat waves 2
historical level
    baselines 129, 134–135, 136–138, 139, 141
    deforestation 140, 141, 143
    policy and institutional reform 219
Houghton, RA 26, 27, 29, 31, 32, 77, 80
HSBC 227
human activities 18–22

IET *see* International Emissions Trading
IFC *see* International Finance Corporation
IIASA *see* International Institute for Applied Systems Analysis
IKONOS 150, 151, 158
illegal logging 35, 45, 90, 237
ILUA *see* Integrated Land Use Assessment
IMAGE model 27
impact assessments 58
impermanence 83, 86, 112, 115, 186
incentives
    cap and trade systems 92, 234
    carbon taxation 91
    deforestation 41, 197
    effective reduction targets 129–143
    emission trading 108
    global policy 35, 238
    long-term framework 119, 121, 122, 123–124, 124–125, 127
    sustainable production 52, 64, 68
income 9, 54, 55, 60, 61–62, 70, 205, 226
Independent Forest Monitors 206
independent voluntary standards 208
India 65, 112, 157

indicative trajectories 130, 142
indicators 217
indigenous communities 9, 43, 45, 55, 195, 205, 229, 230
    see also community level; local level
indirect payments 200
indirect regulation 197
individual cash transfers 191, 197, 200–201, 210
Indonesia 41, 43, 44, 45, 61, 221
industrialised countries 36, 80, 98, 170
industry 18, 87, 174, 181
information transparency 220
infrastructure 43, 49, 58, 194
in-kind payments 197, 200
INPE *see* National Institute for Space Research
institutional level
    capacity 45, 191, 195
    carbon finance 99
    carbon markets 166, 188, 189
    current international framework 102–118
    distribution of finance 200–201
    funding sources 184, 185
    long-term framework 235–236
    measuring and monitoring 162
    reform 211, 213, 214, 216–219, 232, 236, 237
    technology centres 59–60
integrated approaches 28, 54, 83, 95–99, 98–99
Integrated Land Use Assessment (ILUA) 163
intensification of agriculture 53–54, 200
Intergovernmental Panel on Climate Change (IPCC)
    cap and trade systems 98
    deforestation 234
    forest emissions 7
    GHGs stabilization 5, 84
    global temperatures 2
    measuring and monitoring 145, 146, 159, 160, 162, 164
    national inventories 107
    sink effect 21
    SRES 26, 27, 29, 31, 32, 80
international climate change framework 81–118
    current 101–118
International Emissions Trading Mechanism (IET) 108, 109, 113–114
International Environmental Transformation Fund (ETF) 223
International Finance Corporation (IFC) 227
International Finance Facility 228
International Institute for Applied Systems Analysis (IIASA) 75
international level
    action 211–239
    agreements 131
    baselines 135, 143
    carbon markets 127, 166, 172
    credit markets 174, 175–177, 179, 180–181, 189
    distribution of finance 202, 205–208
    ecosystem services 62, 64

international level (*continued*)
  emissions trading 121
  governance and finance distribution 210
  measuring and monitoring 146, 159, 161–162, 163
  mitigation costs 70, 71
  support 70, 71, 205
  sustainable production 68, 239
  transaction log 108, 111, 162
International Transaction Log Administration (ITLA) 111
International Tropical Timber Organisation (ITTO) 217
intra-national leakage 86, 91, 130, 131
inventories
  cap and trade systems 98
  current international framework 104, 105, 115
  governance 206
  Kyoto Protocol 107–108
  measuring and monitoring 112–113, 146, 150, 151–152, 152–155, 162–163, 235
  standards 159
investment
  carbon markets 184, 186, 188–189
  CDM 110, 114
  early action 232
  funding gap 222, 229
  policy and institutional reform 218
  private sector 226
  technology 235
  see also private sector finance; public sector finance
ITLA *see* International Transaction Log Administration
ITTO *see* International Tropical Timber Organisation
Iwokrama, Guyana 226

Japan 170, 171, 174
JI *see* Joint Implementation mechanism
JISC *see* Joint Implementation Supervisory Council
Joint Implementation (JI) mechanism 108, 109, 110, 114
Joint Implementation Supervisory Council (JISC) 111
judicial effectiveness 200

KDP *see* Kecamatan Development Programme
Kecamatan Development Programme (KDP), Indonesia 202, 203–204
Keidanren Voluntary Action Plan 174
Kew report 152–155
knowledge 146, 162, 213, 214–216, 229, 236, 237, 238
  see also technical assistance
Kyoto Protocol
  cap and trade systems 94
  carbon markets 170
  current international framework 103–104, 105–106, 107–116
  emission reduction 169
  national inventories 160, 161
  targets 101

labour 58–59
land
  availability 49, 50–51, 52
  cover 17, 18, 27
Landsat 150, 151, 155
land tenure security
  delegation 199–200
  drivers of deforestation 35, 36, 44–45
  governance 191, 192, 193, 235, 237
land use
  amplifier 21
  avoided deforestation 74
  change emissions 20, 25
  economic returns 41
  measuring and monitoring change 156–157
  planning 193–194, 195, 237
  policies and deforestation 42
land use, land-use change and forestry (LULUCF) 105, 107, 109, 114
  see also agriculture, forestry and other land use
Latin America 27, 46, 72
lCERs *see* long-term Certified Emission Reductions
leakage
  baselines 124, 130, 131, 134, 135, 137, 140, 143
  cap and trade systems 92, 99
  carbon taxation 91
  effectiveness 86, 87, 129, 130
  measuring and monitoring 112
  mitigation 83
  opportunity costs 74
leakage of emissions 86, 87
legal level 35, 43, 45, 193, 201
leisure time 10
liability 122, 124, 127, 129, 130, 224
LIDAR *see* Light Detection and Ranging
Light Detection and Ranging (LIDAR) 151
Lightening the Lode 58
light penetration 20
limited liability 122, 124, 129, 130
linking to carbon markets 165–190, 196, 211, 235
  see also carbon markets
litter 19
livelihoods 9, 52, 220, 237, 238
local level 24, 43, 131, 192, 199
  see also community level; indigenous communities
logging 20, 35, 45, 57, 58, 90, 194, 237
long-term approaches 81–118, 119, 121–127, 234
long-term Certified Emission Reductions (lCERs) 108–109, 110, 115, 186
long-term liability 187

low-carbon development 167, 193, 200
low-cost abatement 69, 70, 189
LULUCF *see* land use, land-use change and forestry

MACCs *see* marginal abatement costs curves
Madagascar 154–155
MAGICC model 32
mainstreaming 192–193, 232
management of finance 201–204
managerial support 200
mandatory standards 207
mapping 145, 147, 149, 151–155, 162, 194, 236
marginal abatement costs curves (MACC) 72–73, 75, 76, 124–125
marginalisation 60, 131, 205
marginal unit of abatement 178–179
marine emissions 228
market aspects 63, 114, 126–127, 165, 213, 223–226, 237
Marrakesh Accords 116
Mayan farmers 55
MDBs *see* Multilateral Development Banks
measuring and monitoring
  baselines 143
  capacity 70
  current international framework 116, 117
  emissions 145–164
  long-term framework 98, 99, 121, 122, 126, 127
  medium-term approach 107–108, 112–113, 119
  transitional systems 236
medium-term approaches 119–210, 234, 238
Merrill Lynch 221
methane 19
Mexico 55
middle classes 37, 39
middle income countries 114, 124
mid-latitude regions 21, 23, 24, 26, 33
migration 43
Millennium Ecosystem Assessment 8, 9
minimum prices 188
mining operations 58
mitigation
  Bali Action Plan 11
  climate change 5–6
  costs 13, 28, 69–80, 95, 142
  deforestation 28, 238
  forest inclusion 234
  leakage 83, 86, 87
  see also abatement
MODIS 150, 151
monitoring 72, 145, 154–155
  see also measuring and monitoring
Montreal Protocol 132
Montserrat 153–154
moral hazards 187–188
Mostizo farmers 55
Mount Oku, Cameroon 152–153
Multilateral Development Banks (MDBs) 231

multilateral funding 95, 196, 213, 222, 229, 230–231, 232, 237
multi-spectral technology 151, 157, 158

national credit registries 235
National Forest Monitoring and Assessment Programme, FAO 163
National Forest Programme Approach 195
National Institute for Space Research (INPE), Brazil 157–158
national level
  baselines 130, 131–132, 133
  capacity building 213
  cap and trade systems 92
  carbon markets 127, 171
  current international framework 86, 105–106
  deforestation 214, 232, 236
  demonstration activities 220
  differentiated emissions 94
  distribution of finance 191, 197–199, 201, 202, 210
  emissions trading 108, 121, 122, 123–124
  governance 191, 192–195, 205, 210, 239
  inventories 98, 104, 105, 107–108, 112–113, 115, 150, 159, 160–161, 235
  long-term framework 236, 238
  measuring and monitoring 145, 146, 159, 160–161, 162
  policy and institutional reform 218
  registries 207
  reversal of emissions 186, 187
  sustainable production 53, 56, 68
  targets 238
  UN Rio Conventions 103
  voluntary standards 208
national sovereignty 89, 92, 106, 108, 131, 192, 205, 207, 208, 234, 235
natural disturbances 88
natural reforestation 23
Near Real Time Deforestation Detection (DETER) 157
negative externalities 85
net benefits 77–80
net deforestation rates 23
net financial flows 98, 99
net global land cover 27
net present value (NPV) 30, 32, 77, 78–79
net primary productivity 17, 18
New Zealand 170, 171, 173
NGOs *see* non-governmental organisations
Nilsson, S 51
nitrous oxides 19
Noel Kempff Mercado Climate Action Project, Bolivia 155
non-Annex I countries
  caps 186
  carbon markets 167, 172
  current international framework 101, 104, 116
  effective reduction targets 129, 130

non-Annex I countries (*continued*)
    Kyoto Protocol 105, 106, 109, 110, 112, 113, 115
    reporting 206
    scaling up 132
    transition pathways 123, 124
    see also developing countries
non-carbon dioxide 18, 181
non-carbon ecosystem services 49, 52
non-carbon market finance 166
non-energy emissions 6
non-forested land 51, 54
non-governmental organizations (NGOs) 199, 224
non-market finance 165, 167
non-state actors 204
non-timber forest production 55–56
Normalized Difference Fraction Index 158
North America 23
Northern hemisphere snow cover 3
Northern Republic of Congo 58
Norway 221, 223
NPV *see* net present value

OA *see* official aid
Oceania, forest area change 23
ODA *see* overseas development assistance
ODI *see* Overseas Development Institute
OECD *see* Organization for Economic Co-operation and Development
off-farm employment 49, 58–60
Office of Climate Change Global Carbon Finance (GLOCAF) 96, 98
official aid (OA) 95
ongoing costs 70–71
opportunity costs 69, 70, 72–74, 76
Organization for Economic Co-operation and Development (OECD) 95
    see also Annex I countries; Annex II countries
overseas development assistance (ODA) 95
Overseas Development Institute (ODI) 217
ownership 43, 44, 199–200
    see also land tenure security
oxidation of soil 19

PAGE model 28
PAMS *see* policies and measures
Papua New Guinea 44
Paris Declaration 230
participation
    baselines 132
    cap and trade systems 92, 99
    community 235
    distribution of finance 191, 192, 195, 201, 202, 203, 210, 237
    forest credit mechanisms 126
    governance 191, 192, 195, 210, 214
    international 86, 89, 211
    Kyoto Protocol 112, 113, 114
    policy and institutional reform 217
    protected areas 60–61
    public funding 229, 230
partnerships 49, 58, 65, 163, 231
past emissions (PE) 135, 137
past interventions 219
payment for ecosystem (environmental) services (PES) 49, 61–62, 72, 196, 198, 199, 220, 221
PE *see* past emissions
peer review 191, 205, 206
PEFC *see* Programme for Endorsement of Forest Certification Schemes
PEN *see* Poverty and Environment Network
performance 130, 137
permanence 86, 88, 91, 92, 99, 186
Peru 45
PES *see* payment for ecosystem (environmental) services
Phase III 177, 178, 179
    see also European Union Emissions Trading Scheme
photosynthesis 18
piloting 224, 237
Pinus spp 147
planning 193–194
plantations 24, 51, 56, 156
Plan Vivo approach 54, 55, 210
PoA *see* Programme of Activities
point scoring 209
policies and measures (PAMS) 206
policy
    baselines 131, 142
    conservation 60–62
    deforestation 192, 197, 205
    distribution of finance 191, 201
    drivers of deforestation 35, 36, 42–43, 48
    emissions reduction 71–72
    governance 191
    incentives 119
    lack of capacity 45
    new international mechanism 193
    reform 211, 213, 214, 216–219, 232, 236, 237
    sustainable production 49, 52, 56, 62, 64
    transition paths 238–239
    voluntary standards 208
political commitment 210
polluter pays principle 92
poor countries 76–77, 145, 161
population 35, 37–38, 47, 48, 49, 50
poverty 7, 9, 49–68, 84, 96, 97–98, 99, 207
Poverty and Environment Network (PEN), CIFOR 215–216
power sector 174, 180, 181
precipitation events 2
pre-determined trajectories 142
preferential procurement 64, 66, 239
preferential treatment 208
premium credits 191, 207, 208, 209
preparatory work 213, 214

prices
  cap and trade systems 93–94
  carbon 90, 98
  carbon markets 165–166, 167, 174–182, 183, 185, 188–189
  CERs 114
  certainty 77
  deforestation 41, 63
  efficiency 89
  food 39, 40
  gas 86
  opportunity costs 73, 74
  stability 235
private sector finance
  carbon markets 165, 167
  distribution 199, 204
  emissions trading 173
  international action 211, 213, 214, 237
  JI 110, 114
  long-term framework 122, 125, 126
  policy and institutional reform 218
  pump-priming 222, 223–227
  responsible purchasing 66
PRODES *see* Annual Deforestation Rate Assessment
production, sustainable 49–68
productivity 17, 18, 35, 39, 49, 51, 53, 59
profits 71, 72
ProForest 217
Programme of Activities (PoA) 132
Programme for Endorsement of Forest Certification Schemes (PEFC) 208
project-based approaches 87, 110, 113, 130, 132, 208
projected baselines 136, 139
projections 15, 26, 37, 75, 77–80, 87, 170
Projeto Ambé, Brazil 57
property rights 44–45
  see also land tenure security
protected areas 49, 60–61, 197, 198, 237
provincial level 199
proximity to roads 145, 155, 158
public awareness 67–68
public sector finance
  carbon markets 167
  distribution 204
  funding gap 222–223, 224
  governance 229–230, 231
  international 211, 237
  long-term framework 90–91, 122, 125, 126
  medium-term 213, 214
pump-priming 213, 222, 223–226, 237
purchasing 75–77, 225, 226

quality control 153
Quantified Emission Limitations or Reduction Commitments 107
Quebec, Canada 147

RADAR *see* Radio Detection and Ranging
Radio Detection and Ranging (RADAR) 151, 153
Rainforest Foundation 194
rainforest regions 36, 46, 217
rapid forest inventory and mapping 152–155
readiness functions 230, 231
REDD *see* reducing emissions from deforestation and degradation
reduced impact logging 56
reducing emissions from deforestation and degradation (REDD) 6–7
  cap and trade systems 83, 96, 97
  carbon markets 168, 183
  demonstration activities 213, 219, 221
  emissions trading 106
  FCPF 230
  Kyoto Protocol 109, 110
  long-term framework 127
  measuring and monitoring 113, 161
  reference levels 127, 129, 130–143, 188
  see also baselines
reforestation 23, 105, 145, 155, 156
  see also afforestation and reforestation
reforms 117, 216–219
Regional Integrated Silvopastoral Ecosystem Management Programme (RISEMP) 54
regional level 122, 123–124, 171, 199
registries 162, 207
regulation
  carbon dioxide 90
  CDM 115
  climate 13, 15, 16, 33
  deforestation 45, 64, 197
  ecosystems 10
  forest credits 117
  sustainable production 64
reinforcing loops of forest transition 47–48
remote sensing 145, 150, 151, 155, 158
removal units (RMUs) 106, 109, 112, 115
renewable resources 209, 228
Renewal Fuels and Fuel Quality Directives 66
rent 48, 76, 77, 183
reporting 105, 191, 205–207
research 145, 147, 159, 211, 213, 214–216, 229, 232, 236
reserves 186, 187–188, 228, 235
resilience 10, 53
responsibility
  common but differentiated 94, 104, 110, 206
  corporate social 66
  delegation of 197–201
  developed countries 122
  taxation 92
  UN Rio Conventions 104
responsible purchasing 66
retention of forest credits 189
retrospective monitoring 154–155
revenue 192, 197, 210, 228

reversal of emissions 184, 186–188, 235
reviews 107, 145, 161–162, 206
rights 9, 35, 193, 195, 228
  see also user rights
Rio Declaration on Environment and
    Development, 1992 104
Rio Earth Summit, 1992 103
RISEMP see Regional Integrated Silvopastoral
    Ecosystem Management Programme
risk 69, 77, 184, 186–189, 205, 222, 224
RMUs see removal units
road building 43, 47
Roberts, D 51
robust measuring and monitoring see measuring
    and monitoring
Royal Botanic Gardens, Kew 152, 155
rules-based trajectories 142
running costs 163
run-off 20
rural nature conservation agreements 61
Russian Federation 170

safeguards 205, 210
sampling 155
Sao Paulo proposal 124
satellite technology 145, 147, 150, 151, 152–153,
    157–158, 164
savings 77
SBI see Subsidiary Body for Implementation
SBSTA see Subsidiary Body for Scientific and
    Technical Assistance
scale 10, 11, 86, 117, 119, 182–184, 232
scaling up 132, 213, 217, 222, 223, 224, 228
SCF see Strategic Climate Fund
science and technology 59–60
Scolel Te, Mexico 55
sea-levels 1, 2, 3
secretariats 110, 213, 229, 230, 237
selective logging 20
self-finance 213, 216, 218, 222
sequestration
  cap and trade systems 97
  carbon 15, 16, 18, 147
  carbon dioxide 24–26, 33
  deforestation and degradation 19
  emissions trading 173
  forest credits 186
  measuring and monitoring 112, 155–159
  national inventories 115
  RMUs 106, 109
  sustainable production 55
  voluntary standards 210
services see ecosystem services
SFM see sustainable forest management
short-term approaches 121, 125, 126, 127, 145,
    211–232, 234, 238
silvopastoral systems 54
sinks 21, 105, 186

slash and burn 20, 26
small countries 163
small-scale aspects 62, 66, 73, 115, 200
snow cover 3
social forest services 10
social impact assessments 58, 191, 192, 205
soil 20, 147, 156
solar radiation 25
soot aerosol emissions 26
South America 23, 43
Southeast Asia 46
special fund mechanisms 201–204
Special Report on Emissions Scenarios (SRES),
    IPCC 27, 29, 31, 32, 80
spending options 196
stabilisation
  baselines 142
  cap and trade systems 96, 97, 99
  carbon dioxide 1, 86, 88
  current international framework 101
  forest transition 47, 48
  GHGs 5–6, 33, 104, 172, 233, 234
  Kyoto Protocol 111–112, 117
  options comparisons 90–95
  resilient forest ecosystems 10
  targets 33, 84, 85, 98, 99, 111–112
  temperature 5, 84
stability 189
standards 66, 159, 187, 191, 205, 207–210, 225
standing forests 129, 130, 134, 136, 138, 140, 143
Staples 67
start-up funds 221
step-change approaches 49, 50, 51, 84
Stern Review 4, 6, 74, 88
stock/average emissions baselines 135–136, 138,
    139, 140–141
stocks 16, 17
Strategic Climate Fund (SCF) 231
stratified sampling 155
sub-national level
  baselines 130–131, 132
  carbon taxation 91
  delegation 192
  distribution of finance 197–199, 203–204
  governance 195
  measuring and monitoring 145, 161
  pump-priming 225
  sustainable production 53
sub-Saharan Africa 53, 222
Subsidiary Body for Implementation (SBI) 110
Subsidiary Body for Scientific and Technical
    Assistance (SBSTA) 110
subsidies 42, 43, 56
subtropical regions 149
supplementarity limits
  EU ETS 165–166, 184
  international credit markets 180, 189
  non-Annex I credits 167, 172

prices 174, 175, 176–177, 178–179
    scale 182, 183, 184
supply and demand 168–174
support
    conservation costs 70
    funding gap 222–223
    infrastructure expansion 50
    international 205, 215, 232, 236–237, 238
    measuring and monitoring 163
    mitigation costs 71, 80
    multilateral funds 230–231
    policy and institutional reform 218
    transparency 206–207
surface water 25
surplus allowances 170
sustainability 66, 199
sustainable development 106, 110, 207
sustainable forest management (SFM) 49, 55–57,
    66–67, 95, 197, 208, 226, 232, 237
sustainable production 13, 49–68, 211, 220, 233,
    238, 239
Switzerland 171, 173

Tapajós, Brazil 57
targets
    carbon markets 174, 177, 178, 182, 183, 184, 189
    current international framework 101, 116, 117
    distribution of finance 199
    effective 129–143, 234, 235, 238
    global temperature 165, 167
    Kyoto Protocol 107, 111–112
    limited liability 122
    long-term framework 121, 125–126, 127, 236
    measuring and monitoring 146
    national 99, 238
    stabilisation 33, 84, 85, 98, 99, 111–112, 233, 234
taxation
    alternative funding sources 228
    deforestation 64
    long-term framework 83, 90, 91, 92, 94, 99
    policy incentives 42, 43, 197
tCERs see temporary Certified Emission
        Reductions
technical level 54, 200, 215
technology
    Bali Action Plan 11
    carbon markets 165
    centres 59–60
    GHGs stabilization 6
    land availability 51
    measuring and monitoring 119, 146, 162, 163
    satellite 145, 147, 150, 151, 152–153, 157–158, 164
    sustainable production 53–54
    transition pathways 235
temperate regions 23, 24–25, 149
temperature
    carbon dioxide 29, 30, 32
    carbon markets 165, 167, 168, 172

changes 1, 2, 3
    impacts 4
    stabilisation 5, 84
temporary Certified Emission Reductions (tCERs)
        108–109, 110, 115–116, 186
terrestrial surface energy balance 25
testing approaches 122, 220
The Nature Conservancy (TNC) 58
tiered approaches 160, 162
timber
    certification 66–67
    deforestation 49, 63, 238
    demand 35, 39–40, 49, 51
    legality assurance 65
    sustainable production 55–56, 57
time-averaged carbon densities 156
timescales 16, 47–48, 129, 133, 141–142
TNC see The Nature Conservancy
Tobin tax 228
tourism 60
tradable emissions see emissions trading
trade-offs 52, 58, 166, 179, 180, 188, 192, 195
traditional rights 193
trajectories, baseline 141–142
transaction costs 72, 88, 101, 115, 117, 196, 200
transition theory 35
transparency
    carbon markets 189
    distribution of finance 191, 201, 202, 203,
        205–207
    information 220
    lack of 45–46
    measuring and monitoring 145, 161, 164
transport 43, 182
triggers of forest transition 47
tropical regions
    carbon dioxide emissions 24, 25
    carbon stocks 149
    carbon storage 17
    deforestation 15, 23, 26
    PEN 215–216
    sink effects 21

UK see United Kingdom
Ukraine 170
Ulu Masen, Indonesia 221
UNCCD see United Nations Convention to
        Combat Desertification
uncertainty
    abatement costs 89
    baselines 130, 131, 136, 137
    carbon dioxide emissions 26, 27, 28
    carbon stocks 24, 149
    future projections 15
    land availability 50, 51, 52
    measuring and monitoring 145, 146, 160, 161
    ownership 199–200
understorey vegetation 158

underwriting private finance 227
undisturbed forests 19
uniform carbon prices 91, 92, 95
United Kingdom (UK) 66, 86, 88, 96, 217, 221, 223, 227
United Nations Convention to Combat Desertification (UNCCD) 103
United Nations Declaration on the Rights of Indigenous Peoples 195
United Nations Framework Convention on Climate Change (UNFCCC)
  carbon markets 168, 170
  current framework 102, 103, 104, 105–106, 107
  GHGs 84, 101
  institutions 110–111
  long-term framework 81, 187
  mandatory standards 207
  measuring monitoring 160–161
  national inventories 115
  opportunity costs 74
  pump-priming 225
  reporting 206
  see also Kyoto Protocol
United Nations (UN)
  distribution of finance 237
  public funding 229, 230, 232
  research and analysis 215
  Rio Conventions 101, 102–106, 117
United States of America (US) 42, 112, 171, 184
up-front finance 70, 71, 222, 226
US *see* United States of America
user rights 44, 191, 192, 193, 235

valuing carbon
  agricultural intensification 54
  climate stabilisation 90
  current international framework 113–114, 116
  externality 60
  sustainable production 13, 52, 62, 63–64
VCM *see* Voluntary Carbon Market
VCS *see* Voluntary Carbon Standard
vegetation mapping 154–155
verification 65, 145, 146, 152, 155–159, 161–162, 207, 235
Verified Emissions Reductions (VERs) 221
VERs *see* Verified Emissions Reductions
village distribution of finance 203–204
virgin forest 42
Voluntary Carbon Market (VCM) 224
Voluntary Carbon Standard (VCS) 187
Voluntary Partnership Agreements (VPAs) 65
voluntary standards 191, 205, 207–210
VPAs *see* Voluntary Partnership Agreements
vulnerability 205

Warner–Leiberman Bill, US 184
waste 8
water 4, 10, 25, 53
wealth 35, 37–38, 48, 49, 50
weighting factors 137, 138
welfare 4
West Africa 46
Wildlife Conservation Society, Congo 58
wood products 9, 10, 156–157
World Bank 163, 203, 206, 217, 225, 229, 231, 237
Worldwide Governance Indicators Index 217

yields 39

zoning programmes 193–194